Crafting Images With Generative Adversarial Networks (GANs) and Models

Parul Dubey
Symbiosis Institute of Technology, Symbiosis International University, Pune, India

Sapna Singh Kshatri
Shri Shankaracharya Institute of Professional Management and Technology, India

Devanand Bhonsle
Shri Shankaracharya Technical Campus, Bhilai, India

Bui Thanh Hung
Industrial University of Ho Chi Minh City, Vietnam

IGI Global
Scientific Publishing
Publishing Tomorrow's Research Today

Published in the United States of America by
IGI Global Scientific Publishing
701 East Chocolate Avenue
Hershey, PA, 17033, USA
Tel: 717-533-8845
Fax: 717-533-8661
Website: https://www.igi-global.com E-mail: cust@igi-global.com

Library of Congress Cataloging-in-Publication Data

Names: Dubey, Parul, editor. | Kshatri, Sapna Singh, 1986- editor. | Bhonsle, Devanand, 1982- editor. | Hung, Bui Thanh, editor.
Title: Crafting images with generative adversarial networks (GANs) and models / edited by Parul Dubey, Sapna Kshatri, Devanand Bhonsle, Hung Bui.
Description: Hershey, PA : IGI Global, [2025] | Includes bibliographical references and index. | Summary: "This book provides readers with a comprehensive understanding of Generative Adversarial Networks (GANs) and empower them to effectively utilize GANs for image generation purposes, from basic principles to advanced techniques"-- Provided by publisher.
Identifiers: LCCN 2024048488 (print) | LCCN 2024048489 (ebook) | ISBN 9798369371527 (hardcover) | ISBN 9798369371534 (paperback) | ISBN 9798369371541 (ebook)
Subjects: LCSH: Image processing--Digital techniques--Case studies. | Generative art--Case studies. | Neural networks (Computer science)--Case studies. | Machine learning--Case studies. | Two-person zero-sum games--Case studies.
Classification: LCC TA1637 .C728 2025 (print) | LCC TA1637 (ebook) | DDC 006.6--dc23/eng/20250105
LC record available at https://lccn.loc.gov/2024048488
LC ebook record available at https://lccn.loc.gov/2024048489

British Cataloguing in Publication Data
A Cataloguing in Publication record for this book is available from the British Library.

Table of Contents

Detailed Table of Contents

estimates them against real data, and these two networks are encouraged to improve iteratively. In particular, this chapter presents you with thorough code snippets in Python using the TensorFlow library, which will lead you through the proper selection and pre-processing of a dataset, the design of generator and discriminator networks, and training.

This paper presents a framework for generating realistic photographs using Generative Adversarial Networks (GANs). Our framework comprises two key neural networks: the Generator and the discriminator. The generator network creates images, while the discriminator network evaluates their authenticity. This framework highlights the potential and advancements of GANs in creating photorealistic visuals. To demonstrate the comparative analysis, we created two distinct models with different picture sizes 128 pixels and 64 pixels, respectively using two different GPUs, the T4x2 and P100. The produced pictures from the P100 GPU have been surpassed by those from the T4x2 GPU, the P100 model has been outperformed by the T4x2 GPU, by considering metrics like LOSS_G, LOSS_D, Real Score, and Fake Score. From the P100 model to the 'T4x2 model, the average value of phony SCORE shows a noteworthy improvement of about 59.34\%, suggesting improved performance in identifying phony data in the later model.

The chapter "Challenges and Future Directions" in the book "Crafting Images with Generative Adversarial Networks (GANs) and Models" delves into the current obstacles and prospective advancements in the realm of GANs and generative models. This chapter aims to provide readers with a thorough understanding of the primary issues hindering the progress and broader application of GANs, while also highlighting the innovative solutions and future directions that could address these challenges. The first section of the chapter explores the major challenges faced by researchers and practitioners in the field of GANs. These include training instability, where GANs often suffer from issues like mode collapse and vanishing gradients, making the training process unpredictable and inefficient.

Generative Adversarial Networks (GANs) have been applied to various fields such as drug discovery, media detection, fashion, autonomous cars, healthcare, natural language processing, and engineering design. To improve unsupervised learning, GANs are made up of a generator network and discriminator that engage in adversarial training. In order to improve the fidelity of data distribution, the generator creates synthetic data that makes it difficult for the discriminator to distinguish between generated and real data. New developments in GAN architectures, such as DCGAN and WGAN, customise models for particular applications, such as hierarchical learning and stable distribution assessment. GANs impact is highlighted by their applications in drug discovery, image synthesis, and style transfer. They are transforming AI in the gaming, cybersecurity, and medical sectors by generating realistic data and encouraging innovation.

This chapter explores the functionalities of Generative Adversarial Networks (GANs), which are employed for the purpose of image synthesis, surpassing conventional methods in their advancements. GANs use a generator and discriminator network for increasing image quality through adverse training. This introduction discusses the progression of methods used for image generation before the advent of GANs. Additionally, it provides an overview of the fundamental principles behind GAN training. While more sophisticated GAN architectures, such as DCGANs, utilize CNN to ensure stability, StyleGAN improves the accuracy of synthesis. Conditional GANs and impact of having domain specific using SRGAN, Text-to-Image to help boost the image quality undergone at major phases. Datasets and pre-trained models such as ImageNet, CelebA, CIFAR correspondingly in significant for training data. The chapter also addresses the application in art, fashion and garment design, medicine for imaging to medical its usage educations media affects creativity and AI training.

In 2014, Ian Goodfellow and colleagues introduced Generative Adversarial Networks (GANs), transforming AI by enabling the creation of highly realistic synthetic data, including text, music, and images. GANs consist of two neural networks: the generator, which creates data samples starting from random noise and improves them iteratively, and the discriminator, which distinguishes between real and generated data, providing feedback to enhance the generator's realism. This adversarial process drives both networks to improve until the generator produces high-quality synthetic data. GANs excel in representing complex data distributions without explicit probabilistic models, ideal for tasks like image generation. However, GAN training is challenging, with issues like training instability and mode collapse. Researchers have developed strategies like regularization techniques and architectural adjustments to address these problems. Despite challenges, GANs are widely used in computer vision, natural language processing, music creation, drug discovery, and privacy preservation.

Generative Adversarial Networks (GANs), introduced by Ian Goodfellow in 2014, have revolutionized machine learning by creating new data. GANs consist of two parts: the Generator, which creates data samples from random noise, and the Discriminator, which distinguishes real from fake data. These networks compete, with the Generator improving its fakes and the Discriminator enhancing its detection. This competition drives both networks to improve. Training GANs can be challenging due to issues like instability and mode collapse. Advanced versions like Conditional GANs (c GANs) and Wasserstein GANs (WGANs) address these problems. This chapter introduces GANs, explaining their components, adversarial training, and practical implementation using the MNIST dataset. It includes guidance on tools, dataset understanding, architecture design, and visualization techniques for evaluating generated images.

Generative Adversarial Networks (GANs) have emerged as versatile tools for synthesizing content across multiple domains. Despite their capability to produce realistic outputs, ongoing research endeavors to further enhance their quality. Key objectives in GAN development encompass reducing training loss and enhancing performance metrics like Inception Score and Fréchet Inception Distance (FID). This paper introduces a novel method to enhance the quality of generated images from textual descriptions by incorporating a secondary loss function. The proposed approach focuses on refining the alignment between images and text during training, aiming to improve the fidelity of synthesized outputs. Leveraging the KT-GAN framework, which extends AttnGAN by integrating attention mechanisms for relevance computation, our method seeks to fine-tune the image-text matching process. Detailed descriptions of the existing KT-GAN system and the proposed enhancements are provided in subsequent sections, demonstrating the potential for significant advancements in GAN-based image synthesis.

This chapter examines the application of Generative Adversarial Networks (GANs) for the recognition and preservation of the Brahmi script, an essential component of cultural and historical heritage. It covers the script's significance, the challenges of ancient script recognition, and introduces GANs, including DCGAN, CycleGAN, and StyleGAN, highlighting their strengths in generating synthetic data, improving image quality, and pattern recognition. The chapter details the methodology of data collection, GAN architecture modifications, and training processes, emphasizing performance metrics like Structural Similarity Index (SSIM), Peak Signal-to-Noise Ratio (PSNR), accuracy, precision, recall, and F1-score. The findings show StyleGANs excel in image quality while DCGANs perform better in accuracy and precision. Practical applications discussed include enhancing historical research, creating digital archives, and developing educational tools. The chapter concludes with future research directions for further improvements in script recognition and preservation.

Preface

Over the last ten years, Generative Adversarial Networks (GANs) have transformed the landscape of artificial intelligence and machine learning, and image synthesis and generation, in particular. These revolutionary deep learning models have shown an ability to generate photorealistic images, improve image resolution, and aid in various creative and industrial uses like never before. Whether it was lifelike virtual avatars or re-imagining how healthcare diagnostics through medical imaging, GANs closed the gap between imagination and reality.

Crafting Images With Generative Adversarial Networks (GANs) and Models has been conceptualized as a reference book for the readers wishing to familiarize themselves with GANs, their incredible capabilities, and their implementations in different areas of research from industrial to academic, beginner to scholars, and professionals. Combining the most recent developments, theoretical teachings, practical methods and real-world examples, the book offers a versatile approach to the topic.

Contents of this Book The chapters of this book cover more or less all aspects from the fundamental principles of GANs to advanced architectures, optimization strategies, and novel applications. Although GANs are most commonly associated with generating highly complex data, several chapters explore specific areas like text-to-image synthesis, transfer learning, photorealism frameworks, and ensemble modeling, demonstrating their versatility in modern research and applications. Others sweep through the incorporation of GANs into creative and pedagogical realms, giving reader's insight into their transformative potential through the redefinition of existing boundaries.

The efforts from different experts from all around the world have come together to present different views, research results, and applications of GANs. From their handover, the diversity of topics speaks to the multi-faceted nature of GANs and how they have been adapted to fields as diverse as healthcare, education, language processing, and more. We would like to thank all contributors who provided their insights and helped in pushing the boundaries of knowledge in this promising and important research field.

We hope that this book serves as a source of inspiration, knowledge, and practical guidance for its readers. This book has something for you, whether you are an experienced researcher pushing the frontiers of GANs or a beginner taking the initial steps into this intriguing domain.We would also like to express our sincere appreciation to IGI Global for the assistance and support throughout this book development. Without them working so hard, this project would not have been possible at all.

Finally, we would like to thank the readers for being interested in this book. With that, we whiz you through the chapters, we hope you will learn and leverage that knowledge of GANs to create you own marvels!

Parul Dubey

Symbiosis Institute of Technology, Symbiosis International University, Pune, India

Sapna Singh Kshatri

Shri Shankaracharya Institute of Professional Management and Technology, India

Devanand Bhonsle

Shri Shankaracharya Technical Campus, Bhilai, India

Bui Thanh Hung

Industrial University of Ho Chi Minh City, Vietnam

Chapter 1
Next-Gen Ophthalmology: Advanced GAN-Based CNN Algorithm for Ophthalmic Abnormality Classification

Siji A. Thomas
Independent Researcher, UAE

Sini Kurien Thomas
Independent Researcher, USA

ABSTRACT

The examination of fundus images using deep learning (DL) technology is crucial for the early identification of retinal abnormalities. The effectiveness of DL algorithms is often limited by the need for large datasets while image fidelity is important in medical imaging, however constrained due to patient's privacy and high costs. This chapter investigates the innovative use of Generative Adversarial Networks (GANs) to address these challenges. Focusing on DC-GAN-based Convolutional Neural Networks (CNNs), the authors focuses the classification of two major retinal abnormalities: Diabetic Retinopathy (DR) and Age-related Macular Degeneration (AMD). Experiments on GAN synthesized retinal images validated the classification of conditions: Normal, DR, or AMD. The developed algorithm achieved impressive performance metrics, with an accuracy of 96.7%and sensitivity of 0.93, while processing images in approximately 4 seconds. Furthermore, the chapter discusses the challenges associated with GANs and suggests potential solutions for future advancements in this field.

DOI: 10.4018/979-8-3693-7152-7.ch001

1. INTRODUCTION

In recent years, retinal images have become tremendously significant in the area of ophthalmology owing to its role in early diagnosis of various ophthalmic abnormalities. Among the various conventional methods, the Deep Learning (DL) can be reasoned as the most emerging technology in classification. Despite its high grade classification ability, the system requires huge datasets for training and validation process. Collecting the medical data is a complicated procedure and may be limited by costs; hence raising the problem of insufficient retinal images. The state-of-art Generative Adversarial Networks (GANs) which highlights the power of artificial intelligence (AI); has indeed the most promising solution for generating synthetic data to address the challenges in computer vision tasks such as image generation & classification, object localization, semantic segmentation etc. GANs leverage the power of artificial intelligence to create realistic synthetic data, reducing the reliance on large datasets for training models. This innovative approach helps in eliminating the need of gathering huge datasets and also enhances the performance and robustness of computer vision systems by providing diverse and augmented data for training. Research studies have indeed explored the performance of Generative Adversarial Networks (GANs) across various imaging modalities, including Magnetic Resonance Imaging (MRI) (Azin *et al*, 2021), Computed Tomography (CT) (Jose *et al*, 2023) and Positron Emission Tomography (PET) (Lei Bi *et al*, 2017), and to name a few. These studies delved the effectiveness of GANs in various tasks such as image generation, denoising, and reconstruction within the medical imaging domain. Presented the article by Goodfellow (2016), mentioned that GAN network has the ability to handle sharp estimated density functions, to efficiently generate desired samples, the elimination of deterministic bias, and has good compatibility with the neural architectures. By leveraging GAN technology, researchers aim to enhance the quality of medical imaging data, improves diagnostic accuracy, and facilitate advancements in healthcare through innovative applications of artificial intelligence. GANs can be used to generate synthetic images of normal and abnormal medical conditions. By training GANs on normal medical images, any deviation from the generated normal images can be flagged as a potential anomaly, aiding the detection of some abnormalities in the medical image. The authors in this chapter, aims to employ Deep Convolutional GAN-based CNN architecture for classification, by generating synthetic data for training, sequentially reducing the computational complexity which in turn helps in better classification.

GAN is a neural network where the networks, namely Generator and the Discriminator performs a tandem competition like a minimax game-scenario. GANs exploit the back- propagation without complex markov-chain. Every iteration begins with the Generator synthesizing the input, which in modern applications it would be typically an image. Thus, the Generator tries to generate the data with same features of the training dataset which seems the real data. Next, the adversarial network; also called Discriminator (act like a classifier); uses it as input to classify whether the generated image is real or fake; if the latter, the Generator's parameters are adjusted with the aim of generating an output which is as realistic as possible. The figure 1 illustrates the flow diagram of a GAN architecture.

Figure 1. Flow diagram of a GAN architecture

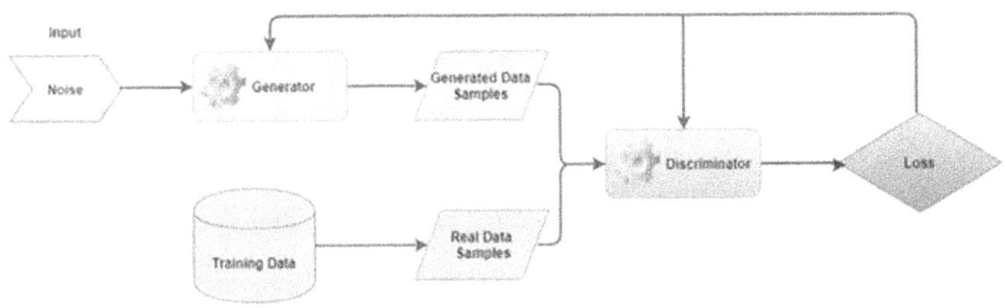

This chapter mainly focus to distinguish between normal retinal images with the prominent ophthalmic abnormalities such as Diabetic Retinopathy (DR) and Age-related Macular Degeneration (AMD). DR is the latter stage of diabetes mellitus, and if not diagnosed and treated properly, the retinal blood vessel gradually damage, resulting a condition of total blindness. AMD is characterized by the appearance of a progressive degenerative change in the macular area of retina. Figure 2 demonstrates the fundus image of normal retina, and DR & AMD affected retinal images.

Figure 2. Normal, DR and AMD retinal images

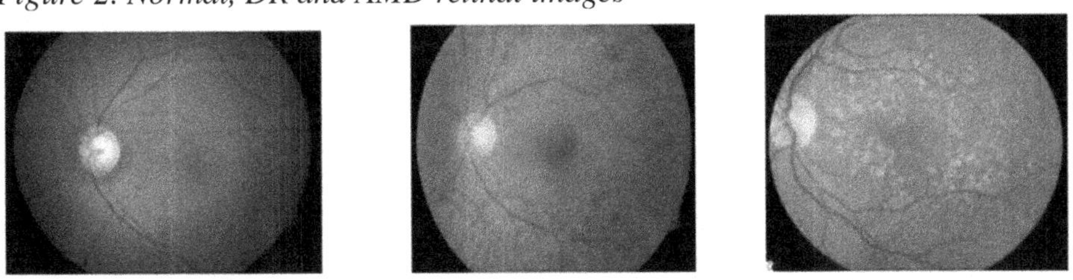

Many DL algorithms have been developed over the past years to detect DR (Gulshan *et al* (2016), Abramoff *et al* (2016), Ting *et al* (2017), Gargeya *et al* (2017) and AMD (Ting *et al* (2017), Hood *et al* (2018)). Authors in this chapter, employs the Convolutional Neural Networks (CNN), a promising technology which is dedicated mainly for classification purpose; a class of deep learning algorithm.

CNN directly learn features from the raw images during the training, eliminating the need of manual feature extraction and hence performs the classification tasks. The image is given to CNN layers followed by the pre-processing step which involves the process of rescaling. The image is rescaled before it is applied to the ConvNet in order to get a fixed number of feature values for all the images that are trained. The convolutional, rectified linear activation function (ReLU) and the pooling layer performs the feature learning and the combination of these layers re-iterates. A pile of convolutional and pooling layers learn the features deeply and are processing in 2D plane. It then moves to the classification tasks, which are performed by fully-connected and the softmax layer. In this algorithm, the authors have added local response normalization, average pooling and dropout layer in the higher levels of CNN to deeply classify the feature values. The general architecture of CNN is illustrated in figure 3.

Figure 3. General architecture of CNN with its basic layers

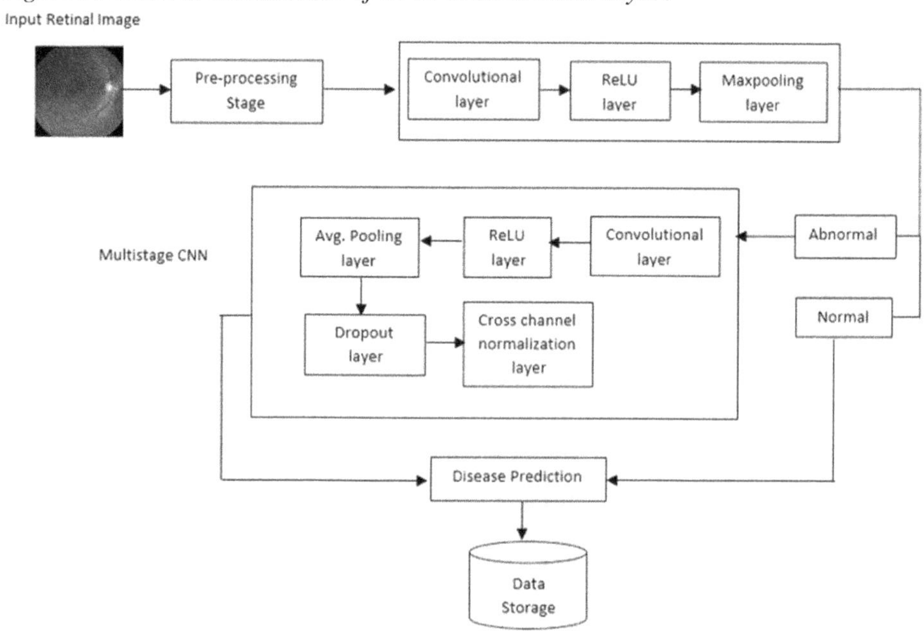

GANs are "the most interesting idea in the past decade in the field of machine learning," according to Yann LeCun (Arthur *et al,* 2014), who co-founded the modern deep neural network revolution with Geoffrey Hinton. The authors focused on to the problem of insufficient medical images for various applications, thus proposed GANs to enhance the performance in classifying the retinal disease. The GAN-based CNN architecture can be considered as a high-grade classification technique owing the following:

- Convolutional filters can deeply extracts the subtle features from the raw images
- Ability to generate high resolution images
- Reduces analytical complexity
- High quality performance comparing the other state-of-the-art techniques

This chapter delves with the core principle of GAN-based CNN algorithm implemented for diagnosing the retinal abnormalities and discusses both the networks developed. The chapter also highlights its technical key challenges and outlines the suggestions for future research.

2. BACKGROUND

Due to the remarkable advancement in computer vision tasks, the applications of neural networks have attracted a lot of interest in medical imaging community in the past recent years. However, the lack of widespread adoption of machine learning software in clinical practice is primarily attributable to the reality that neural networks can met errors when conditions like adaptation issues or missing data occurs. A possible explanation is significantly its smaller size of fully-annotated medical imaging datasets compared to other domains. In contrast to a typical medical image dataset, which is three to four orders of magnitude smaller, the gold standard computer vision ImageNet (Deng *et al,* 2009) dataset has over 14 million annotated images. This is because, compiling medical imaging datasets is expensive and challenging, since accurate annotation of the sensitive data necessitates a very specific domain knowledge base. The lack of training data in medical imaging has prompted researchers to actively explore alternative methods to obtain the training sets (Frangi *et al*, 2018). Thus, GANs had become the standard approach to generate synthetic images from real data.

2.1 GAN Related Works

The use of GANs in medical imaging goes beyond image synthesis. The review done by Kazeminia *et al*, 2020 researched on: synthesis, segmentation, reconstruction, detection, de-noising, registration, and classification in medical image processing and discusses the performance score of various architectures such as: MRI (Magnetic Resonance Imaging), CT (Computed Tomography), OCT (Optical

Coherence Tomography), chest X-ray, dermoscopy, Ultrasound, PET (Positron Emission Tomography) and microscopy. The researchers in Iqbal *et al,* 2018 developed a GAN architecture to generate synthetic medical images and its segmented masks. The model achieved 0.837 and 0.832 dice coefficient in STARE and DRIVE dataset respectively. In Francesco *et al,* 2023, the authors developed a new model to discriminate the generated real-fake retinal images, the algorithm achieved the F- measure of 0.95. Andreini *et al,* 2021 developed a two stage GAN for generating high quality retinal images, the semantic label-maps are generated describing the blood vessel structure. The authors in Y. Salini *et al*, 2022 developed a cycle GAN network, which focuses on the generation and segmentation of retinal images. The performance were measured using images from DRIVE and the algorithm achieved 98.19% accuracy. Burlina with other researchers (2019) created ProGAN models to generate intermediate and/or advanced AMD images where they achieved an accuracy of ~54%. Researches in Youssef *et al*, 2023 presented some top performing GANs, the authors tested the algorithm on various medical imaging modalities and achieved good performance rates.

Lee *et al*, 2021 developed a cGAN model to predict the presence of four abnormal structures on post treatment OCT using a larger dataset of 15,183 paired OCT B-scans from 241 patients to predict post treatment OCT images in patients with Neovascular AMD receiving anti-VEGF. After including images from indocyanine green angiography and fluorescein angiography (FA) in the training datasets, the predictive performance improved. The authors led to the conclusion that the model is unsuitable for using it as a screening tool due to its low sensitivity.

Table 1. Shows few existing work on GANs for synthesizing the coloured retinal fundus images.

References	**Datasets**	**Methods**	**Result (Accuracy)**
Dereje *et al* (2022)	DRIVE	PU-Net 33	0.98
Kamran *et al* (2021)	DRIVE, CHASE-DBI, STARE	RV-GAN	0.95
Guibas *et al* (2017)	MESSIDOR	cGAN	0.88
Iqbal *et al* (2018)	STARE, DRIVE	MI-GAN	0.83

2.2 CNN Related Works

For the past few years, the researchers have shown a growing interest in the application of CNN for abnormality detection and disease classification (Fryan *et al*, 2023, Alaba *et al*, 2023, Chan *et al*, 2020). In 2021, Mohamed *et al* developed a CNN algorithm for automatic retinal abnormality detection. The authors discusses the classification of ocular abnormalities in retinal images, and achieved good performance score. Harry Pratt *et al*, 2016 mainly focussed on the detection of DR along with its severity stages like mild, moderate, severe and proliferative. Harry and his co-authors separated augmented image followed by pre-processing and used 10 convolutional layers to learn more features deeply with three fully connected layers. The authors trained 10,290 images with adopted filters of 3*3 window size. The CNN algorithm was trained on fundus images available at Kaggle dataset using Graphical Processing Unit (GPU) and achieved a high sensitivity of 0.95 with an accuracy of 75%. Carson et al introduced an automatic screening system for the DR identification by adopting a CNN (Carson Lam *et al*, 2018). They trained and tested on Google Net and Alex Net. Experiments were conducted on Kaggle dataset and Messidor-1 datasets. The pre- processing involves Otsu's method which performs cropping and Contrast Limited Adaptive Histogram Equalization (CLAHE) for the contrast adjustments. The transfer learning scenario was adopted instead of Fully-connected layer. By the usage of CLAHE pre- processing, the CNNs ability to extract the subtle features improved and they achieved an accuracy of 74.5%, 68.8% and 57.2% for 2-aray, 3-aray and 4-aray respectively. Felix *et al*, 2018 proposed a deep learning classification system for detecting the severity scale of AMD, utilizing 13 classes of AMD with an ensemble of deep learning architectures, which in turn improved the accuracy. The authors measured the outcome based on kappa statistics. Subsequent to the training stage, an evaluation metric compares the CNN output iteratively with the true class thereby adjusting the weights to minimize the loss. The algorithm employs random forest for classification purpose. Authors in Chao Chen *et al*, 2025 surveyed and detailed the overview of main algorithms utilized in medical image classification. The research also addresses the evolution of CNN for medical image classification, the research challenges in clinical application, and the future research directions. Govindaiah *et al*, 2018 studied on deep convolution neural networks for finding the relevance of using the transfer learning to screen an individual at risk of Age-related Macular Degeneration (AMD) using Age-Related Eye Disease Study (AREDS) dataset with more than 150000 images and achieved an accuracy of around 92% with modified VGG16 network.

3. METHODOLOGY

The dataset was analysed and processed for two different classes (training and validation) which contains 462 retinal images of DR, AMD and Normal. The images were collected from ARIA (2006), STARE (2004) and E-OPHTHA (2006) where 300 images were used for training and 162 for testing and 162 for validation purpose. Table 2 shows the samples taken for each class in training and validation set.

Table 2. Retinal disease samples before using GAN

Category	Training	Validation
Diabetic Retinopathy	*100*	*53*
Age related macular degeneration	*100*	*53*
Normal	*100*	*56*
Total	*300*	*162*

Despite the high classification accuracy achieved accuracy of deep learning technology, data imbalance negatively impacts the performance rates; as cited in F. J. Pulgar *et al*, 2017. To address this issue, samples were evenly distributed in terms of size, colour, and .tiff format. Authors in this chapter put forward the use of a DC-GAN network to enhance the dataset of ophthalmic abnormality images by generating a synthetic dataset.

3.1 DC-GAN Model

Training the GAN architecture is a pivotal process, which involves two distinct networks, the Generator (G) and Discriminator (D), and are competing against each other. The generator attempts to spawn images that seem the real ones; while the discriminator to distinguish between real and the fake fundus images. Random noise is provided as input to the generator, the generated synthetic images merged with the real ones to form a training cluster. The generator output is then passed on to the next neural model- the Discriminator. The loss is calculated to determine the classification accuracy. The weights G(z) are adjusted and updated using gradient descent optimization algorithm to minimize its loss. The Discriminator feeds real and the synthetic data, acting as a classifier. The Discriminator must be trained to get D(x) close to 1 and D(G(z)) close to 0.

The objective function of GAN can be calculated by employing the equation 1.

$$\text{Min}_G \text{Max}_D V(D, G)$$

$$= E_{x \sim pdata(x)} [\log (D(x)] + E_{z \sim pz(z)} [\log (1 - D(G(z)))] \tag{1}$$

where;

G(z)- the output of the Generator;
G- the synthetic data;

D(x)- the output of the Discriminator, representing the probability that x could be 0 or 1

The authors in this chapter uses DC-GAN for the application, which is same as GAN, but it incorporates deep CNN models in both G and D with minor changes in the structure of CNN so as to enhance the quality level and the convergence rate of samples. Figure 4 shows G- network in DC-GAN architecture (Goodfellow *et al*, 2014). The authors made the following changes:

- Use of fractional-stride convolution in the generator and stride convolution for the the discriminator, which affects the size of output feature maps
- Incorporate batch norm in both G and D
- Avoid FC layer
- Apply ReLU activation function for G layer which employs sigmoid activation and leaky ReLU activation function in D layer, which small returns negative values for negative inputs. The activation function helps to learn complex patterns and features.

Figure 4. G- Network in DC-GAN architecture

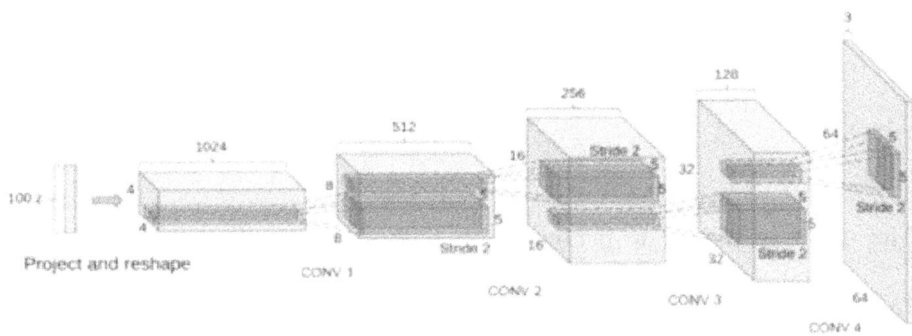

The generated retinal output images from the DC-GAN model is then applied to a detailed CNN architecture to classify the retinal stages; explained in section 3.2.

3.2 CNN Model

An input layer, feature learning layer is created in the first stage CNN, the fully connected layer and the softmax layer performs classification and detection. Feature learning layer includes the convolution, ReLU and the Maxpooling layer. In this chapter, the images that are trained are of size 300∗400∗3, thus the image is of width 300 and height 400 with three colour channels R, G, B. The convolution layer in the CNN algorithm automatically chooses a filter, where the filter size can be defined in the algorithm. The filter slides over the image which in turn performs the convolution and gets a feature map. Briefly, if x is the small part of the image, W the weight, b the bias parameter; the total number of features can be estimated as given in equation 2.

$$W^{T}.\, x + b \qquad (2)$$

ReLU layer follows the max (0,x) concept and does not make any changes in the image size, where x represents the pixel value. ReLU layer makes all negative values to zero, the maxpooling layer performs the down-sampling operation and the size of the image gets shrinks. This layer helps to reduce the computational complexity as well as to reduce the number of parameters. The parameters such as Stride (S) and Padding (P) can be defined at Maxpooling layer. The convolved feature map depends upon three parameters such as number of filters, stride and the padding parameters. However, it is not mandatory to pad the input image always. If the size of the image is $W1 * H1 * D1$, then after passing through the convolution layer, the size would be $W2 * H2 * D2$, where W2, H2 and D2 can be defined by equations 3, 4 and 5.

$$W2 = \frac{W1 - F + 2P}{S} + 1 \qquad (3)$$

$$H2 = \frac{H1 - F + 2P}{S} + 1 \qquad (4)$$

$$D2 = K \qquad (5)$$

The activation maps now becomes, 296* 396* 20 since the size of the Kernel, K is 20. Hence, parameters are generated by equation (6) in the first Convolutional layer by the process of parameter sharing.

$$(F * F * D1).K) + \sum_{i=1}^{20} bi \qquad (6)$$

where, bi is the bias parameter for 20 filters. The number of Convolution, ReLU and Maxpooling layer added, defines the shrinkage of the image. Then the network moves to the classification layer where actual classification occurs in the Fully-connected and the Softmax layer. The Fully-connected layer performs the flattening process. Likewise, these process occurs for all the images that are to be trained. The databases for the first CNN is created for these three categories. The multistage CNN shoots if and only when the sample is an abnormal one. On following the ReLU layer in multistage CNN, Cross channel normalization layer is added to perform channel wise local response normalization. In the multistage CNN, the layers such as Average pooling, Cross channel normalization and the drop out layer is also added.

Figure 5. Architecture for CNN model

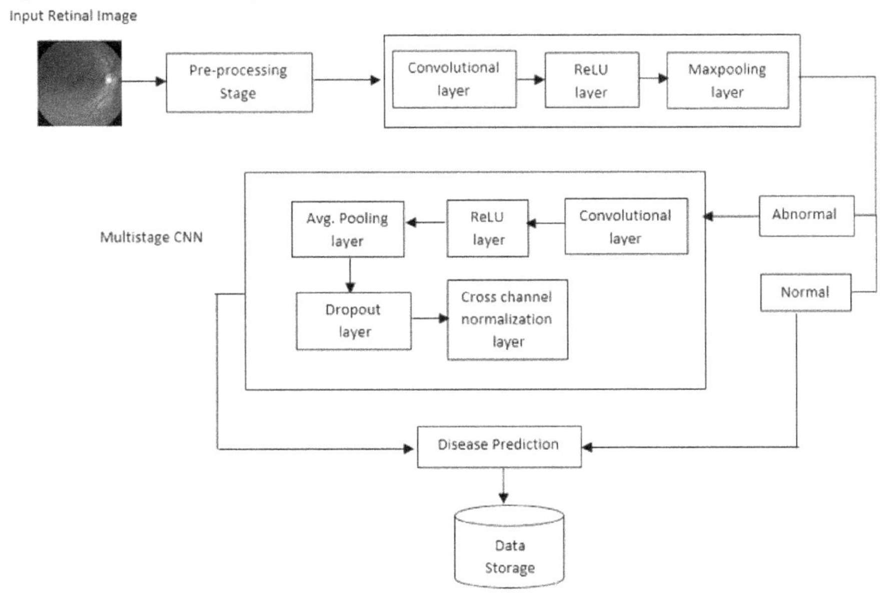

The algorithm classifies retinal image into normal, DR or AMD. If abnormal, it triggers into the next stage of CNN where it performs the higher feature classification, the architecture is as shown in figure 5.

4. EXPERIMENTAL ANALYSIS AND RESULTS

The authors in this chapter aimed to enhance the number of training dataset using DC-GAN architecture, the output of this network feds to the multistage CNN model to improve the ophthalmic disease classification accuracy. The performance of neural networks relies on how the network architecture is built. The results obtained were put through the performance analysis and is described below. Experimental evaluations were done using MATLAB 2017b version in CORE i3 Windows OS.

Table 3 depicts the number of retinal images generated using DC-GAN.

Table 3. Retinal disease samples after using DC-GAN

Category	Generated data using DC-GAN	CNN Model	
		Total Training Data	Validation
Diabetic Retinopathy	*112*	*212*	*53*
Age related macular degeneration	*145*	*245*	*53*
Normal	*133*	*233*	*56*
Total	*390*	*690*	*162*

For the CNN model, the first stage CNN has 10 layers trained with 50 epochs, the second has 12 layers with 80 epochs and third has 15 layers. The CNN stage is trained for 100 epochs to enhance the performance rate. For each stage, the CNN achieved mini-batch loss less than 0.005 with its mini batch accuracy of 100%, the base learning rate fixed to 0.0001. The mini batch accuracy increases as the iterations count increases. All these above mentioned parameter selection makes the system to get good performance scores in accuracy and sensitivity.

The processing performance for GAN and GAN based CNN algorithm were measured separately as referenced in Table 4. The developed system achieved an overall accuracy rate of 96.78% and sensitivity of 0.933.

Table 4. Performance metrics

Metric	Elucidation
Accuracy	TP+TN/ (TP+TN+FP+FN)
Sensitivity	TP/(TP+FN)

The different features learnt by the network can be seen through "deep dream image" and is shown in figure 6.

Figure 6. Deep dream image showing the features learnt by the network in max-pooling layer with 15 channels

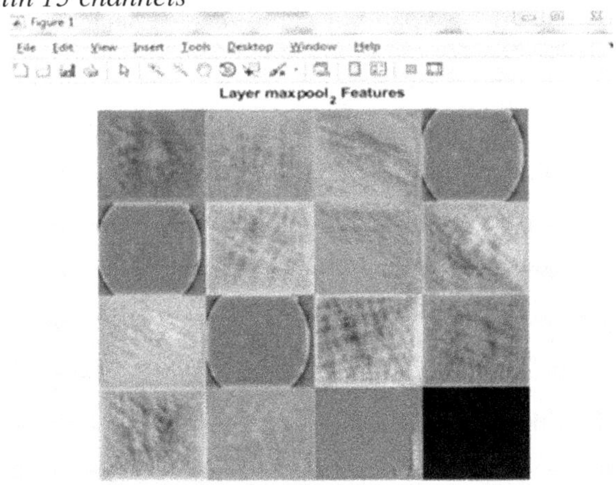

Table 5. Validations on STARE dataset

Model	No. of Epochs	Accuracy	Sensitivity
Performance of DC-GAN Model	50	84.83%	0.86
	100	92.86%	0.94
Classification Performance of DC-GAN based CNN	50	89.67%	0.88
	100	96.78%	0.93

5. DISCUSSION AND CONCLUSION

The demand for artificial biomedical data generation is significant and tremendously grown over the past decade, the authors in this chapter deployed a deep convolutional generative adversarial network (DC-GAN) to generate training images with a multistage CNN model to classify the retinal abnormalities (DR and AMD) with the normal one. The introduction of DC-GAN-based CNN model enhanced the quality of performance and achieved an accuracy of 96.78% with 0.93 sensitivity. The state-of-art, DC-GANs offers significant improvement in the field of medical imaging to generate high quality images to generate the training datasets. The multistage CNN offer several advantages, and performs the accurate classification of various retinal abnormalities. Despite the advantages of GANs, such as providing a cost effective solution and the reduction in extended time process, its training process can be challenging, say due to the absence of defined convergence state, since the two networks compete with each other. Additionally, when the discriminator performs significantly better than a generator, the loss of the discriminator jumps to zero, which causes the retro propagated gradient to get smaller and smaller, the condition so called vanishing gradient, as GANs train a generator using the discriminator's output. Hence; leading to suboptimal results due to inadequacy in the gradient updates. In light of this book chapter, the authors suggests to tailor the usage of a denoising encoder for further studies to improve the accuracy, and more training stability can be yielded by introducing a differentiable augmentation prior the generated images enters the discriminator. Additionally, inception score and frechet inception distance (FID) can be utilized to assess the quality of synthetic images; such as fidelity, functionality, and to name a few.

REFERENCES

Abramoff, M. D., Lou, Y., Erginay, A., Clarida, W., Amelon, R., Folk, J. C., & Niemeijer, M. (2016). Improved Automated Detection of Diabetic Retinopathy on a Publicly Available Dataset Through Integration of Deep Learning. *Investigative Ophthalmology & Visual Science*, *57*(13), 5200–5206. DOI: 10.1167/iovs.16-19964 PMID: 27701631

Akil, M., Elloumi, Y., & Kachouri, R. (2021). Detection of retinal abnormalities in fundus image using CNN deep learning networks. In *State of the Art in Neural Networks and their Applications* (pp. 19–61). Academic Press.

Al Fryan, L. H., Shomo, M. I., & Alazzam, M. B. (2023). Application of Deep Learning System Technology in Identification of Women's Breast Cancer. *Medicina*, *59*(3), 487. DOI: 10.3390/medicina59030487 PMID: 36984487

Alaba, S. (2023). Image Classification using Different Machine Learning Techniques. *TechRxiv*. 1-5.

Andreini, P., Ciano, G., Bonechi, S., Graziani, C., Lachi, V., Mecocci, A., Sodi, A., Scarselli, F., & Bianchini, M. (2021). A Two Stage GAN for High Resolution Retinal Image Generation and Segmentation. *Electronics (Basel)*, *11*(1), 60. DOI: 10.3390/electronics11010060

Miller, A. I. (2019). 10 Ian Goodfellow's generative adversarial networks: AI learns to imagine.

Azin S, David C. R and Viktor V. (2021). CNNs and GANs in MRI-based Cross Modality Medical Image Estimation. *Electrical Engineering and Systems Science: Image and Video Processing,* 1, 01-40.

Bi, L., Kim, J., Kumar, A., Feng, D. D. F., & Fulham, M. (2017). Synthesis of Positron Emission Tomography (PET) Images via Multi-channel Generative Adversarial Networks (GANs). *Lecture Notes in Computer Science*, *10555*, 43–51. DOI: 10.1007/978-3-319-67564-0_5

Burlina, P. M., Joshi, N., Pacheco, K. D., Liu, T. Y. A., & Bressler, N. M. (2019). Assessment of Deep Generative Models for High Resolution Synthetic Retinal Image Generation of Age Related Macular Degeneration. *JAMA Ophthalmology*, *137*(3), 258–264. DOI: 10.1001/jamaophthalmol.2018.6156 PMID: 30629091

Chan, H, Samala, R.K, & Hadjiiski, L.M.; Zhou. (2020). Deep Learning Models in Medical Image Analysis. *Deep Learning in Medical Image Analysis Challenges and Applications*, *1213*, 3–21. PMID: 32030660

Chen, C., Nor, A. M. I., & Liu, X. (2025). A Review of Convolutional Neural Network Based Methods for Medical Image Classification. *Computers in Biology and Medicine*, *185*, 185. DOI: 10.1016/j.compbiomed.2024.109507 PMID: 39631108

Database, A. R. I. A. (2006). Retinal image archive. Available at www.eyecharity.com/aria online.html

Database, S. T. A. R. E. (2004). Structure Analysis of the Retina. Available at www.ces.clemson.edu/ahoover/stare/

Deng, J., Dong, W., Socher, R., Li, L. J., Li, K., & Fei-Fei, L. Imagenet (2009). A Large-scale Hierarchical Image Database; Proceedings of IEEE Conference on Computer Vision and Pattern Recognition. 248–255.

E-Ophtha Database. (2006). A colour fundus image database. Available at www.adcis.net/en/Download-Third-Party/E-Ophtha.html

Frangi, A. F., Tsaftaris, S. A., & Prince, J. L. (2018). Simulation and Synthesis in Medical Imaging. *IEEE Transactions on Medical Imaging*, *37*(3), 673–679. DOI: 10.1109/TMI.2018.2800298 PMID: 30253380

Gargeya, R., & Leng, T. (2017). Automated Identification of Diabetic Retinopathy Using Deep Learning. *Ophthalmology*, *124*(7), 962–969. DOI: 10.1016/j.ophtha.2017.02.008 PMID: 28359545

Goodfellow, L. (2016). NIPS 2016 Tutorial: Generative Adversarial Networks, Arxiv: *Computer Science- Machine Learning*. 6, 01-57.

Govindaiah, A., Hussain, M. A., Smith, R. T., & Bhuiyan, A. (2018, April). Deep convolutional neural network based screening and assessment of age-related macular degeneration from fundus images. In 2018 IEEE 15th International symposium on biomedical imaging (ISBI 2018) (pp. 1525-1528). IEEE.

Grassmann, F., Mengelkamp, J., Brandl, C., Harsch, S., Zimmermann, M. E., Linkohr, B., Peters, A., Heid, I. M., Palm, C., & Weber, B. H. F. (2018). A Deep Learning Algorithm for Prediction of Age Related Eye Disease Study Severity Scale for Age-Related Macular Degeneration from Colour Fundus Photography. *Ophthalmology*, *125*(9), 1410–1420. DOI: 10.1016/j.ophtha.2018.02.037 PMID: 29653860

Guibas, J.T., Virdi, T.S., Li, P.S (2018). Synthetic Medical Images from Dual Generative Adversarial Networks. *Computer Vision and Pattern Recognition*. 3, 01-09.

Gulshan, V., Peng, L., Coram, M., Stumpe, M. C., Wu, D., Narayanaswamy, A., Venugopalan, S., Widner, K., Madams, T., Cuadros, J., Kim, R., Raman, R., Nelson, P. C., Mega, J. L., & Webster, D. R. (2016). Development and Validation of a Deep Learning Algorithm for Detection of Diabetic Retinopathy in Retinal Fundus Photographs. *Journal of the American Medical Association, 316*(22), 2402–2410. DOI: 10.1001/jama.2016.17216 PMID: 27898976

Hood, D. C., & De Moraes, C. G. (2018). Efficacy of a Deep Learning System for Detecting Glaucomatous Optic Neuropathy based on Colour Fundus Photographs. *National Library of Medicine: Ophthalmology, 125*, 1207–1208. PMID: 30032794

Ian, G., Pouget-Abadie, J., Mirza, M., Xu, B., Warde-Farley, D., Ozair, S., Courville, A., & Bengio, Y. (2020). Generative Adversarial Networks. *Communications of the ACM, 63*(11), 139–144. DOI: 10.1145/3422622

Iqbal, T., Ali, H. (2018). Generative Adversarial Network for Medical Images (MI-GAN). Journal of Medical Systems. 42(11), 01-19.

José, M., Tania, P., Francisco, S., Julieta, F., Morgado, J., Freitas, C., Eduardo, N., Beatriz Flor, L., Miguel, C., António, J. M., Isabel, R., José, L. C., Venceslau, H., António, C., & Hélder, P. Oliveira. (2023). Lung CT image synthesis using GANs. *Expert Systems with Applications, Elsevier,* 215, 119350.

Kamran, S. A., Hossain, K. F., Tavakkoli, A., Zuckerbrod, S. L., Sanders, K. M., & Baker, S. A. (2021). RV-GAN: Segmenting Retinal Vascular Structure in Fundus Photographs Using a Novel Multi-scale Generative Adversarial Network. *Lecture Notes in Computer Science, 12908*, 34–44. DOI: 10.1007/978-3-030-87237-3_4

Kazeminia S., Baur C., Kuijper A., Van Ginneken B., Navab N., Albarqouni S., Mukhopadhyay A (2020). GANs for Medical Image Analysis. *Artificial Intelligence in Medical Imaging*.109.

Lam, C., Yi, D., Guo, M., & Lindsey, T. (2018). Automated Detection of Diabetic Retinopathy Using Deep Learning. *AMIA Joint Summits on Translational Science Proceedings AMIA Summit on Translational Science, 17*, 147–155. PMID: 29888061

Lee, H., Kim, S., Kim, M. A., Chung, H., & Kim, H. C. (2021). Post-treatment Prediction of Optical Coherence Tomography Using a Conditional Generative Adversarial Network in Age-Related Macular Degeneration. *Retina (Philadelphia, Pa.), 41*(3), 572–580. DOI: 10.1097/IAE.0000000000002898 PMID: 32568984

Mercaldo, F., Brunese, L., Martinelli, F., Santone, A., & Cesarelli, M. (2023). Generative Adversarial Networks in Retinal Image Classification. *Applied Sciences (Basel, Switzerland), 13*(18), 10433. DOI: 10.3390/app131810433

Pratt, H., & Coenen, F., Deborah M Broadbent, Simon P Harding (2016). Convolutional Neural Networks for Diabetic Retinopathy. *International Conference on Medical Imaging Understanding and Analysis, Science Direct*. 90, 200-205.

Pulgar, F. J., Rivera, A. J., Charte, F., & Del Jesus, M. J. (2017). On the Impact of Imbalanced Data in Convolutional Neural Networks Performance. *Proceedings in International Conference Hybrid Artificial Intelligence System*. 10334, 220–232. DOI: 10.1007/978-3-319-59650-1_19

Shenkut, D., & Bhagavatula, V. (2022). Fundus GAN - GAN-based Fundus Image Synthesis for Training Retinal Image Classifiers. *44th Annual International Conference of the IEEE Engineering in Medicine & Biology Society (EMBC)*. 2185-2189. DOI: 10.1109/EMBC48229.2022.9871771

Skandarani, Y., Jodoin, P. M., & Lalande, A. (2023). GANs for Medical Image Synthesis: An Empirical Study. *Journal of Imaging*, *9*(3), 69. DOI: 10.3390/jimaging9030069 PMID: 36976120

Ting, D. S. W., Cheung, C. Y.-L., Lim, G., Tan, G. S. W., Quang, N. D., Gan, A., Hamzah, H., Garcia-Franco, R., San Yeo, I. Y., Lee, S. Y., Wong, E. Y. M., Sabanayagam, C., Baskaran, M., Ibrahim, F., Tan, N. C., Finkelstein, E. A., Lamoureux, E. L., Wong, I. Y., Bressler, N. M., & Wong, T. Y. (2017). Development and Validation of a Deep Learning System for Diabetic Retinopathy and Related Eye Diseases Using Retinal Images from Multi-Ethnic Populations with Diabetes. *Journal of the American Medical Association*, *318*(22), 2211–2223. DOI: 10.1001/jama.2017.18152 PMID: 29234807

Yalamanchili Salini, J HariKiran. (2022). Deepfakes on Retinal Images using GAN. *International Journal of Advanced Computer Science and Applications*, *18*(8).

Chapter 2
Building Your First GAN Model

Rajendra Kumar Pandey
Shri Shankaracharya Institute of Professional Management and Technology, Raipur, India

Rupali Vyas
https://orcid.org/0009-0006-1382-1141
Shri Shankaracharya Institute of Professional Management and Technology, Raipur, India

ABSTRACT

The advent of generative adversarial networks (GANs) has been a genuine game changer for artificial intelligence and deep learning. It enables computers to generate imaginative and realistic data samples. This chapter provides step-by-step guidance on how to design your very first GAN model. We will cover all elements, theoretical and practical. We shall start with the principles of GAN, followed by mathematical formulations of it and the training process involved in its adversarial character. As you progress through the foundations, you will grasp how GANs effectively have a generator and a discriminator. One develops synthetic data samples, while another estimates them against real data, and these two networks are encouraged to improve iteratively. In particular, this chapter presents you with thorough code snippets in Python using the TensorFlow library, which will lead you through the proper selection and pre-processing of a dataset, the design of generator and discriminator networks, and training.

DOI: 10.4018/979-8-3693-7152-7.ch002

1. INTRODUCTION

GANs have been the most disruptive framework in deep learning, particularly in the area of generative modeling, where the purpose is to synthesize data that looks like a given dataset. First proposed by Ian Goodfellow and his colleagues in 2014(I. J. Goodfellow et al., 2014a; Reed et al., 2016; Torfi et al., 2022), GAN has recently emerged as one of the major techniques to generate high-fidelity synthetic data relevant to practically all areas, such as image production, video synthesis, or even data augmentation(Figueira & Vaz, 2022; König et al., 2023). A GAN will thus have a fundamental architecture of two neural networks: generator and discriminator; all these play diverse, antagonistic roles inside the same model(I. Goodfellow et al., 2020; I. J. Goodfellow et al., 2014b; O'Shea & Nash, 2015).

A generator network is taught in such a way that it can generate synthetic data. Normally, it is from a random noise vector(König et al., 2023). The purpose is to create data that cannot be discriminated against by a discriminator from genuine data, thereby "fooling" the discriminator. On the other side, there is a discriminator operating as a binary classifier, differentiating the real data samples from a training dataset from the synthetic samples made by the generator(Figueira & Vaz, 2022; Reed et al., 2016). The adversarial aspect of GANs then becomes clear since the generator is trained to maximize the error on the discriminator, while the discriminator itself is trained to decrease it(I. Goodfellow et al., 2020; Karras et al., n.d.).

This technique then produces a dynamic and iterative training process in which both networks continuously enhance each other.

The present chapter is focused on the building and training of a GAN model from both conceptual and practical aspects. We will start with very fundamental basic principles regarding GANs, in which mathematical formulations establish an adversarial training process. Understanding this theoretical basis will help you fully understand how the generator generator and discriminator discriminator operate in the GAN system. For specifics in the model architecture discussed in this chapter, see ***Figure 1*** for an understanding of the basic structure of a Generative Adversarial Network (GAN). It is a well-elucidated diagrammatic representation that goes a long way to disclose its primary components involved and how they are used to interact with one another to aid further digest the GAN framework described in this chapter.

Figure 1. Illustration of the basic architecture of a Generative Adversarial Network (GAN) model.

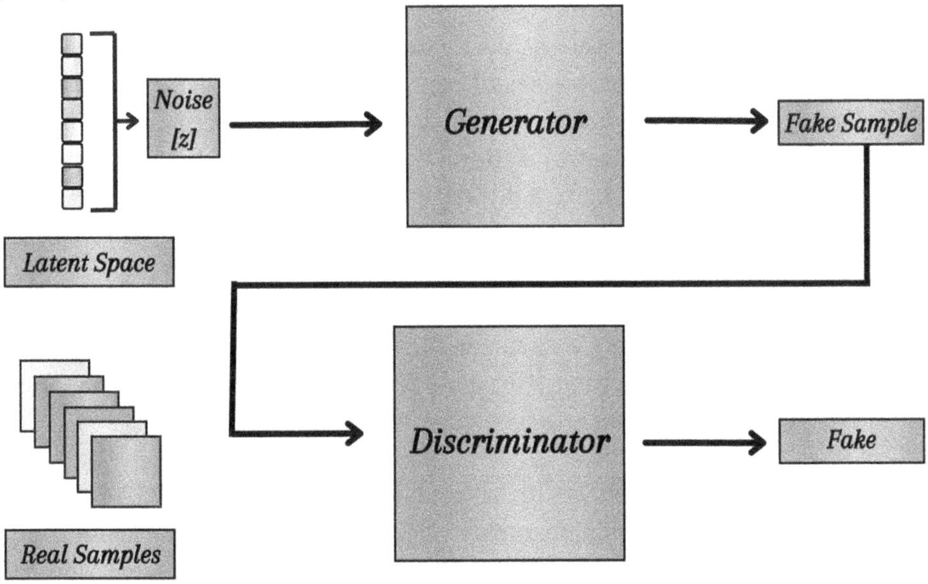

The practical component of this chapter is a line-by-line implementation of a GAN. You will be guided through the means to create the generator and discriminator component network networks; attention will be given to their architectural choices and the roles each perform inside a GAN. The rationale driving deployments has been integrations that assist in numerous crucial decisions linked to GAN training.

A. **Training the Generator and Discriminator:** The common code has functions for training both of these networks at the same time. It provides a *train_step* function that will compute gradients for each network and use those to update the model parameters so that the generator is better at creating realistic data and the discriminator is better at differentiating between them.
B. **Image Generation and Visualization:** A function like *generate_and_save_images* allows one to generate and visualize synthetic images throughout the training process. This notation generates images but may be further conditioned under particular circumstances by human input, something that would go a long way in establishing that, indeed, the GAN is able to create outcomes on demand.
C. **Comparing with Real Data:** The *generate_comparison_image* function is crucial in comparing the created picture against real samples, therefore providing visual evidence on how far the GAN has gone or advanced in generating realistic data. Such a comparison will be critical for measuring the generator's

performance and gaining knowledge of how similar the created data will be to the real dataset.

D. **Training Result Visualization:** Throughout training, the code maps a record of the losses for both the generator and the discriminator, using the plot losses function. This loss plot helps to identify some common training difficulties mode collapse or vanishing gradients and modify the training process accordingly.
E. **User Interaction and Customization:** The sample additionally interacts with the user to create any specified digit upon runtime input. This highlights the dynamic nature of GANs in generating any desired output that a conditional input may demand.
F. **Model saving and final generation:** at the end of training, the models will have to be saved for future use, creating final images that illustrate the GAN's capabilities after full training. This code ensures all training operations, from scratch to final output, are efficient and effective.

This chapter, therefore, aids in understanding and implementing GANs on the basis of both theoretical foundations and practical methodologies. Upon completion of the chapter, you will have obtained an in-depth knowledge base in the working processes of GANs, endowed with the wherewithal, at the same time, to develop, train, and configure your own GAN models for applications spanning subject domains.

2. MATHEMATICAL FOUNDATIONS

The very roots of GANs reside in game theory; the training process can simply be viewed as a two-player minimax game(Luedtke et al., 2020). Mathematical formulations play an important role in how GANs work because they describe what is literally an adversarial connection between the generator and discriminator(Zühlke & Kudenko, 2024).

The generator (G) transfers a random noise vector, generally sampled from a conventional normal distribution, to a data sample (z), which aims to imitate real data from a particular distribution. On the other hand, the discriminator (D) considers a sample (x), either from the real dataset or produced by (G), and produces the probability ($D(x)$) of (x) really being a real sample(Dash et al., 2024).

The adversarial aspect of GANs is mirrored in the cost functions applied to the generator and discriminator, both of which are optimized against each other. Mathematically, this interaction materializes as a game known as a minimax game:

$$min_G max_D V(D,G) = E_{x \sim p_data(x)}[logD(x)] + E_{z \sim p_z(z)}[log(1 - D(G(z)))] \quad (1)$$

Here, $(V(F, G))$ represents the value function that the discriminator (D) and generator (G) aim to optimize. The first term, $\left(E_{x \sim p_data(x)}[logD(x) \right.$ reflects the discriminator's objective to maximize the probability of correctly identifying real samples (x) from the true data distribution $(p_{data}(x))$. The second term, $\left(E_{z \sim p_z(z)}[log(1 - D(G(z)))] \right)$, represents the generator's objective to minimize the probability that the discriminator correctly identifies its generated samples $(G(z))$ as fake(Chakraborty et al., 2024).

2.1 The Optimization Process is Iterative and Adversarial

- The discriminator is trained to maximize its ability to distinguish real samples from fake ones by increasing the term $(logD(x))$ for real samples and $(log(1-D(G(z))))$ for generated samples.
- The generator is trained to minimize the discriminator's success rate by reducing the term $(log(1-D(G(z))))$, which corresponds to "fooling" the discriminator into classifying generated samples as real.

The past decade has seen the evolution of GANs, increasing the spectrum of applications while also enriching our understanding of generative models with unique architectures and training approaches that really push the bounds on what is feasible to be achieved by deep learning. GANs have become a significant building block within the AI research community; every additional sophisticated model advance continues to broaden the potential of generative modeling(Chakraborty et al., 2024; Dash et al., 2024).

It will frame the chapter on GAN designs, the actual mathematical underpinnings of their training methods, and relatively new advancements in the field. Pointers to cutting-edge research shall augment the following explanation, proving that GANs have been in rapid evolution ever since and continue to be important for modern AI today.

3. GAN Model Construction

In this section, we are going to get started on how to design your very first generative adversarial network model from scratch. A GAN, as previously established, comprises two major components: a generator and a discriminator. These elements have various roles within the antagonistic framework. The building of such neural networks is a vital aspect of developing a functional GAN; its success and efficiency depend on the design and interaction between them(Figueira & Vaz, 2022; Karras et al., n.d.; Zhou et al., 2023).

First, we'll define the design of the generator network. Since it is an artistic power engine it allows GANs to create synthetic data samples that look alike with respect to the characteristics of real data the generator ingests a random noise vector, usually from a normal standard distribution, and tries to transform this noise into a data sample similar to those found in the training dataset. The generator's design may vary, but it generally consists of numerous layers of transposed convolutions, also known as deconvolution, that up sample this noise vector into a structured output(Figueira & Vaz, 2022; I. Goodfellow et al., 2020; Karras et al., n.d.).

Now, we will design the discriminator network, operating as a critic in an adversarial interaction. The discriminator would analyse samples and label them with a real or fake score, e.g., if they had been seen in the training dataset or were generated by a generator, respectively. Typical designs for the discriminator frequently resemble a conventional convolutional neural network using some layers of convolutions, down sampling the input data, and eventually ending in a single output neuron expressing the probability that the input is real(Huang et al., 2017; Sengupta et al., 2020).

Once both networks are formed, we integrate them into a cohesive GAN model. As described earlier, this integration defines the loss functions for the generator and discriminator according to the Minimax game. From the generator side, one would want to decrease how well the discriminator is capable of discerning between bogus and real data. From the discriminator side, it will be excellent to maximize its capacity for creating the right differentiation. This interplay of loss functions is what drives adversarial training.

This will be followed by a training process, which is the really essential learning part of GAN to come out with real, indistinguishable data. The technique incorporates an iterative training procedure in which both networks are updated, create data, and eventually modify the parameters for both networks based on their respective losses.

Now, this process is repeated for a number of epochs until the generator is good at generating data that the discriminator can't separate from the real data.

Apart from constructing and training the GAN, this section will also address all other crucial relevant topics:

A. **Data pre-processing:** Before training the GAN with data, it is crucial to pre-process them suitably. That will entail standardizing the data within an acceptable range and reshaping it to have the necessary shape to pass as input into the networks, probably with augmentation done to increase the variety in the training set(Yuan et al., 2023).

B. **Model Evaluation:** The performance of GAN has to be verified during the training phase. This is achieved by showing the generated photographs at regular intervals and checking them visually. This is also performed by plotting the loss curves for the generator and discriminator to monitor their convergence. Overcome Common Pitfalls in Training GANs: Training a GAN can be tricky due to a variety of difficulties, including mode collapse where the generator generates relatively limited variations of outputs or unstable training dynamics(Zhou et al., 2023). 'We will examine strategies to address these issues by modifying the learning rate, regularization techniques, and employing various loss functions, among others, such as the Wasserstein loss(Li et al., 2024).

For a thorough grasp of the procedure presented in this chapter, go to ***Figure 2***. This illustration depicts the full workflow of the generative model, highlighting the movement through several stages of the model architecture to the final output, and so providing a thorough view of the entire process.

Finally, we will end this part by storing the trained models and illustrating how to use them for sample creation. By the end of the section, you will have a fully functional GAN model, ready to explore further and improve upon. The purpose of this section is to educate you not just with the technical abilities required to build a GAN but also more fundamentally grasp what's happening that actually makes GANs such an important tool inside deep learning. Now, we're going to start step by step with developing your GAN model.

Figure 2. An illustration showing the overall workflow of the generative model, through various stages of the model architecture to the final output.

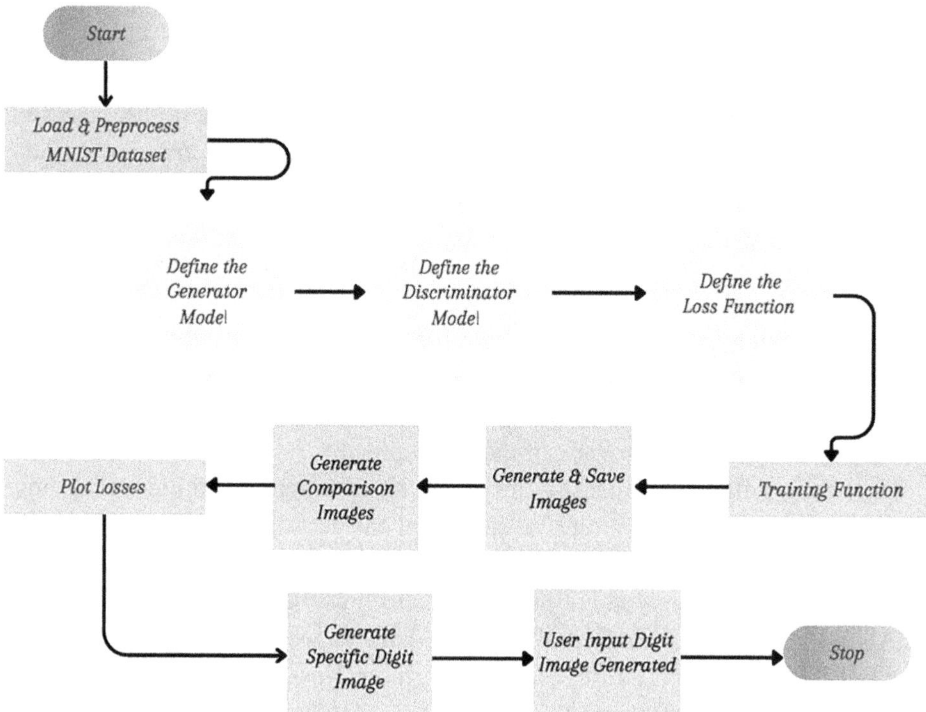

3.1 Data Preparation

First, we want to prepare the MNIST dataset, which contains 28 × 28 grayscale images of handwritten digits(Lecun et al., 1998). We would like to use in-built methods inside TensorFlow for loading this dataset(*Training a Neural Network on MNIST with Keras | TensorFlow Datasets*, n.d.), where the images are normalized to be within the range [-1, 1] so that training for a GAN is stable(Ali et al., 2024). Likewise, labels referring to these digits inside the images are loaded and modified to meet the proportions predicted by our model as input. the data utilized in this investigation, ***Figure 3*** exhibits representative images from the MNIST dataset together with their related labels.

Figure 3. Sample images of the MINST dataset with label.

It is then organized into a TensorFlow data pipeline, shuffling the dataset and batching it into smaller groups of images for efficient processing during training. ***Figure 4*** gives a code snippet that displays the preprocessing of the MNIST dataset for GAN training.

Figure 4. This code snippet prepares the MNIST dataset for GAN training

```
import tensorflow as tf
from tensorflow.keras import layers, Input, Model
import matplotlib.pyplot as plt
import numpy as np
from IPython import display
import time

# Load and preprocess the MNIST dataset
(train_images, train_labels), (_, _) = tf.keras.datasets.mnist.load_data()
train_images = train_images.reshape(train_images.shape[0], 28, 28, 1).astype('float32')
train_images = (train_images - 127.5) / 127.5  # Normalize to [-1, 1]
train_labels = train_labels.reshape(-1, 1)

BUFFER_SIZE = 60000
BATCH_SIZE = 256

train_dataset = tf.data.Dataset.from_tensor_slices((train_images, train_labels)).shuffle(BUFFER_SIZE).batch(BATCH_SIZE)
```

3.2 Model

In a neural network, the generator accepts as input a random noise vector along with a label describing the digit to be created. It generates the output of a synthetic 28×28 image. The generator progressively up-scales its noise vector with dense layers in its architecture, utilizing batch normalization and transpose convolutional layering. Batch normalization will stabilize this training process by normalizing each layer's outputs. Leaky ReLU activations introduce non-linearity and consequently allow the model to learn more complicated patterns(Karras et al., n.d.).

The label is encoded in a vector and multiplied with the noise vector to generate a combined input supplied to the network. This will ensure that the generator is conditioned on the label and thus able to emit visuals corresponding to a given digit. ***Figure 5*** demonstrates a code snippet that implements the Generator model within a GAN framework.

Figure 5. This code snippet is the implementation of the generator model in a GAN

```
# Generator Model
def build_generator():
    noise = Input(shape=(100,))
    label = Input(shape=(1,), dtype='int32')
    label_embedding = layers.Embedding(10, 100)(label)
    label_embedding = layers.Flatten()(label_embedding)

    model_input = layers.multiply([noise, label_embedding])

    x = layers.Dense(7*7*256, use_bias=False)(model_input)
    x = layers.BatchNormalization()(x)
    x = layers.LeakyReLU()(x)

    x = layers.Reshape((7, 7, 256))(x)
    x = layers.Conv2DTranspose(128, (5, 5), strides=(1, 1), padding='same', use_bias=False)(x)
    x = layers.BatchNormalization()(x)
    x = layers.LeakyReLU()(x)

    x = layers.Conv2DTranspose(64, (5, 5), strides=(2, 2), padding='same', use_bias=False)(x)
    x = layers.BatchNormalization()(x)
    x = layers.LeakyReLU()(x)

    x = layers.Conv2DTranspose(32, (5, 5), strides=(2, 2), padding='same', use_bias=False)(x)
    x = layers.BatchNormalization()(x)
    x = layers.LeakyReLU()(x)

    x = layers.Conv2DTranspose(1, (5, 5), strides=(1, 1), padding='same', use_bias=False, activation='tanh')(x)

    model = Model([noise, label], x)
    return model

generator = build_generator()
```

3.3 Discriminator Model

It is a neural network that operates as a binary classifier, distinguishing between actual images collected from the dataset and fake images created by the generator. The discriminator takes an image together with its accompanying label, concatenates them, uses them as input for processing, and produces an evaluation of the likelihood that such an image is real.

The design of the discriminator contains convolutional layers for feature extraction, followed by dropout layers. These layers prevent the model from overfitting the training data by randomly changing a fraction of input units to zero while training. Lastly, the output from the discriminator is a single scalar number, reflecting the likelihood that the input image is real. ***Figure 6*** displays a code snippet that illustrates the implementation of the Discriminator model in a GAN.

Figure 6. This code snippet presents the implementation of the Discriminator model in a GAN

```
# Discriminator Model
def build_discriminator():
    image = Input(shape=(28, 28, 1))
    label = Input(shape=(1,), dtype='int32')

    label_embedding = layers.Embedding(10, 28*28)(label)
    label_embedding = layers.Flatten()(label_embedding)
    label_embedding = layers.Reshape((28, 28, 1))(label_embedding)

    concatenated = layers.concatenate([image, label_embedding])

    x = layers.Conv2D(64, (5, 5), strides=(2, 2), padding='same')(concatenated)
    x = layers.LeakyReLU()(x)
    x = layers.Dropout(0.3)(x)

    x = layers.Conv2D(128, (5, 5), strides=(2, 2), padding='same')(x)
    x = layers.LeakyReLU()(x)
    x = layers.Dropout(0.3)(x)

    x = layers.Conv2D(256, (5, 5), strides=(2, 2), padding='same')(x)
    x = layers.LeakyReLU()(x)
    x = layers.Dropout(0.3)(x)

    x = layers.Flatten()(x)
    x = layers.Dense(1)(x)

    model = Model([image, label], x)
    return model

discriminator = build_discriminator()
```

3.4 Loss Functions and Optimizers

Now, we are going to train the GAN by defining a loss function for the generator and discriminator. The objective would be to decrease the probability of correct classification of the created image by the discriminator on the generator's side, and on the other side, maximize the discriminator based on its distinction between actual and fake images. We will use binary cross-entropy loss in both networks because it is particularly appropriate for binary classification tasks. In this paper, an Adam optimizer is utilized for the weights' optimization corresponding to each of these two neural networks. It is more efficient and ensures an adaptable learning rate(Dash et al., 2024; Karras et al., n.d.). ***Figure 7*** specifies the loss functions and optimizers applied for training both the Generator and Discriminator models in a GAN. This code sample illustrates the particular criteria and strategies utilized to maximize the performance of these models, ensuring effective adversarial training.

Figure 7. This code snippet defines the loss functions and optimizers used for training the generator and discriminator models in a GAN.

```
# Loss and Optimizers
cross_entropy = tf.keras.losses.BinaryCrossentropy(from_logits=True)

def discriminator_loss(real_output, fake_output):
    real_loss = cross_entropy(tf.ones_like(real_output), real_output)
    fake_loss = cross_entropy(tf.zeros_like(fake_output), fake_output)
    return real_loss + fake_loss

def generator_loss(fake_output):
    return cross_entropy(tf.ones_like(fake_output), fake_output)

generator_optimizer = tf.keras.optimizers.Adam(1e-4)
discriminator_optimizer = tf.keras.optimizers.Adam(1e-4)
```

3.5 Training Process

This will become a series of alternating updates to the generator and discriminator. The primary actions involved in each training step are as follows:

Noise and Label Generation: The generator's input is a random batch of noise vectors with accompanying labels.
Synthetic picture generation: It employs noise and labels to generate synthetic images.
Discriminator Evaluation: Real images from the dataset and synthetic images from the generator are sent through the discriminator. It computes the losses for both the generator and discriminator and adjusts their weights via back-propagation.

The method is performed for a number of epochs set by the user, regenerating and saving images at regular intervals to illustrate the progress towards the GAN. ***Figure 8*** depicts the training loop for the GAN, demonstrating the iterative process involved in updating both the Generator and Discriminator models. This code snippet provides a full picture of how the models are trained across numerous iterations, including the procedures used to refine their performance.

Figure 8. This code snippet outlines the training loop for the GAN, detailing the iterative process of updating both the Generator and Discriminator models.

```python
# Training
@tf.function
def train_step(images, labels):
    noise = tf.random.normal([BATCH_SIZE, 100])
    generated_labels = tf.random.uniform([BATCH_SIZE, 1], minval=0, maxval=10, dtype=tf.int32)

    with tf.GradientTape() as gen_tape, tf.GradientTape() as disc_tape:
        generated_images = generator([noise, generated_labels], training=True)

        real_output = discriminator([images, labels], training=True)
        fake_output = discriminator([generated_images, generated_labels], training=True)

        gen_loss = generator_loss(fake_output)
        disc_loss = discriminator_loss(real_output, fake_output)

    gradients_of_generator = gen_tape.gradient(gen_loss, generator.trainable_variables)
    gradients_of_discriminator = disc_tape.gradient(disc_loss, discriminator.trainable_variables)

    generator_optimizer.apply_gradients(zip(gradients_of_generator, generator.trainable_variables))
    discriminator_optimizer.apply_gradients(zip(gradients_of_discriminator, discriminator.trainable_variables))
```

3.6 Monitoring and Saving Progress

During training, the progress of a GAN should be tracked by producing and saving images at regular intervals. In this manner, we can visualize the quality of the images made by the generator and be confident that everything is going smoothly. Additionally, the models are saved after being trained so that additional experimental work or employing them in other applications for trial will not be restricted. Figure 9 explains how to monitor and preserve the training progress of the GAN. This code sample offers methods for setting checkpoints and displaying generated images at various stages, offering insights into the model's evolution and performance throughout the training process.

Figure 9. This code snippet demonstrates how to monitor and save the training progress of the GAN, including checkpoints and visualizations of generated images at various stages.

```python
def generate_and_save_images(model, epoch, test_input, test_labels, save=False):
    test_labels = tf.convert_to_tensor(test_labels, dtype=tf.int32)  # Ensure test_labels are tensors
    predictions = model([test_input, test_labels], training=False)

    fig = plt.figure(figsize=(4, 4))

    for i in range(predictions.shape[0]):
        plt.subplot(4, 4, i+1)
        plt.imshow(predictions[i, :, :, 0] * 127.5 + 127.5, cmap='gray')
        plt.axis('off')

    if save:
        plt.savefig('image_at_epoch_{:04d}.png'.format(epoch))
    plt.show()

def generate_comparison_image(model, epoch, real_images, real_labels, save=False):
    real_labels = tf.convert_to_tensor(real_labels, dtype=tf.int32)  # Ensure real_labels are tensors
    noise = tf.random.normal([real_images.shape[0], 100])
    generated_images = model([noise, real_labels], training=False)

    fig, axs = plt.subplots(2, real_images.shape[0], figsize=(20, 4))

    for i in range(real_images.shape[0]):
        axs[0, i].imshow(real_images[i, :, :, 0] * 127.5 + 127.5, cmap='gray')
        axs[0, i].axis('off')
        axs[1, i].imshow(generated_images[i, :, :, 0] * 127.5 + 127.5, cmap='gray')
        axs[1, i].axis('off')

    if save:
        plt.savefig('comparison_image_at_epoch_{:04d}.png'.format(epoch))
    plt.show()

seed = tf.random.normal([16, 100])
seed_labels = np.array([num for _ in range(2) for num in range(8)]).reshape(-1, 1)

def train(dataset, epochs):
    gen_losses = []
    disc_losses = []

    for epoch in range(epochs):
        start = time.time()

        for image_batch, label_batch in dataset:
            train_step(image_batch, label_batch)

        # Generate and save images
        display.clear_output(wait=True)
        generate_and_save_images(generator, epoch + 1, seed, seed_labels, save=True)

        # Generate comparison images
        sample_images = train_images[:8]
        sample_labels = train_labels[:8]
        generate_comparison_image(generator, epoch + 1, sample_images, sample_labels, save=True)

        # Compute losses for visualization
        noise = tf.random.normal([BATCH_SIZE, 100])
        generated_labels = tf.random.uniform([BATCH_SIZE, 1], minval=0, maxval=10, dtype=tf.int32)
        generated_images = generator([noise, generated_labels], training=False)
        real_output = discriminator([train_images[:BATCH_SIZE], train_labels[:BATCH_SIZE]], training=False)
        fake_output = discriminator([generated_images, generated_labels], training=False)
        gen_loss = generator_loss(fake_output)
        disc_loss = discriminator_loss(real_output, fake_output)

        gen_losses.append(gen_loss)
        disc_losses.append(disc_loss)

        print(f'Time for epoch {epoch + 1} is {time.time()-start} sec')
        print(f'Generator loss: {gen_loss}, Discriminator loss: {disc_loss}')

    # Generate after the final epoch
    display.clear_output(wait=True)
    generate_and_save_images(generator, epochs, seed, seed_labels, save=True)
    generate_comparison_image(generator, epochs, sample_images, sample_labels, save=True)

    # Plot the losses
    plot_losses(gen_losses, disc_losses)
```

3.7 Visualizing and Saving Results

After training, there is a requirement to check GAN's performance by producing images and graphing the loss curves. It also depicts generator and discriminator losses in terms of epochs to help understand the dynamics of training. Moreover, one can retain trained models for usage later on, where they can further be fine-tuned or employed in other future projects. ***Figure 10*** shows Plotting generated results and loss metrics The accompanying code snippet offers the ability to visualize the quality of the generated output and what progress is taking place in training using these loss curves for both models: Generator and Discriminator.

Figure 10. This code snippet show the process of visualizing the generated results from the GAN and the plot losses.

```
def plot_losses(gen_losses, disc_losses):
    plt.figure(figsize=(10, 5))
    plt.plot(gen_losses, label='Generator Loss')
    plt.plot(disc_losses, label='Discriminator Loss')
    plt.xlabel('Epochs')
    plt.ylabel('Loss')
    plt.legend()
    plt.show()

# Function to generate an image of a specific digit
def generate_digit_image(digit):
    noise = tf.random.normal([1, 100])
    digit_label = tf.convert_to_tensor([[digit]], dtype=tf.int32)
    generated_image = generator([noise, digit_label], training=False)

    plt.figure(figsize=(2, 2))
    plt.imshow(generated_image[0, :, :, 0] * 127.5 + 127.5, cmap='gray')
    plt.axis('off')
    plt.show()
```

3.8 Training and Test User Input

Finally, the GAN model is trained for the specified number of epochs. After training, we would have a trained generator for creating new images based on user input. This would give an example of how a GAN generates defined-digit pictures from a given label. ***Figure 11*** shows the training phases and evaluation procedures for the trained model. This code sample shows how the model is examined using user input and image production tasks post-training, giving a methodology for evaluating the effectiveness and performance of the GAN beyond the training phase.

Figure 11. This code snippet outlines the training steps and evaluation of trained model used to assess the user input and image generation after training.

```
# Train the model
EPOCHS = 100
train(train_dataset, EPOCHS)

# Generate an image of a specific digit (e.g., digit 3)
generate_digit_image(3)

# Ask the user to input a number
user_input = int(input("Please enter a digit (0-9): "))

# Call the function with the user-provided digit
generate_digit_image(user_input)
```

4. RESULTS AND EVALUATION

The GAN model trained on the MNIST dataset could generate remarkably realistic images of the digits. In this section, the specifics concerning the performance of the GAN model, the obstacles in training it, and the qualitative and quantitative evaluation linked to the generated images will be given.

4.1 Loss and Training Progress Analysis

Mostly, throughout the course of this training, generator losses increased linearly, as did discriminator losses. The generator loss dropped while the model made its ease of generation better in a bid to generate more realistic images that may trick a specifically developed discriminator.

In contrast, the discriminator loss decreased early when it learned to identify reals from synthesized images. Then, as the generator got better at delivering high-quality imagess, it stabilized over time. This balance of generator and discriminator losses is critical because significant imbalances are likely to cause mode collapses or vanishing gradients.

The graph of the generator and discriminator losses over training epochs headed toward convergence, indicating that both networks achieved some stable point at which the generator provided enough realism in images while the discriminator was correct regarding real versus fake images. Figure 12 is a graphic illustrating the Generator and Discriminator loss during GAN training from epoch 0 to 100. This animation emphasizes the dynamic learning process as the models iteratively update their outputs, displaying how the losses evolve over time and providing insights into the convergence behavior of the GAN.

Figure 12. Plot showing the generator and discriminator loss during GAN training from epoch 0 to 100, highlighting the dynamic learning process as the models iteratively refine their outputs.

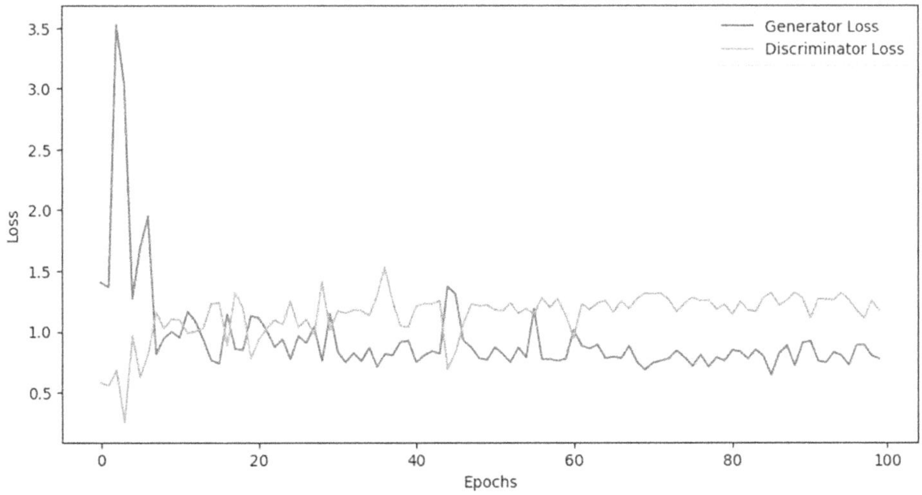

4.2 Image Quality of Generated Images

The quality of the generated images improved progressively as the training epochs increased. In the early epochs, the images generated by the GAN were quite fuzzy and did not have a clear digit structure. However, around epoch 50, the images seem to be something like recognized digits with properties that distinctly correlate to different classes of the MNIST dataset. ***Figures 13, 14, and 15*** demonstrate the training sample images in different epochs when training a GAN. In ***Figure 13***, one can recognize a sample image from epoch 1 of the GAN training process; it shows an object view of how the model began to output initially. Then, ***Figure 14*** gives a sample image at epoch 50 to present intermediate progress and, lastly, ***Figure 15*** delivers a sample image at epoch 100 to give a view of the refined result of the model after extensive training.

Figure 13. Training sample image at epoch 1

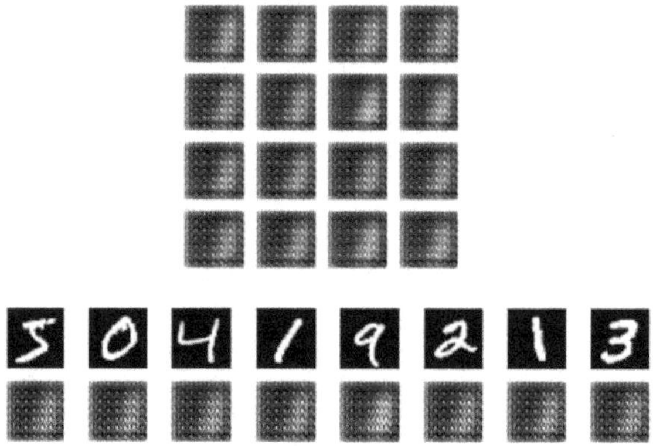

Qualitatively, the created images were viewed by sampling and viewing some of the generations across distinct epochs. The images created after 100 epochs were practically indistinguishable from the original MNIST digits, with very sharp edges and a clear outline of the digit. In this conditional GAN configuration, the generator may now be conditioned to generate images depending on specified digit labels. The generator did quite effectively output digits corresponding to the given labels; this proved that the model could learn the relationship between noise input and the label.

Figure 14. Training sample image at epoch 50

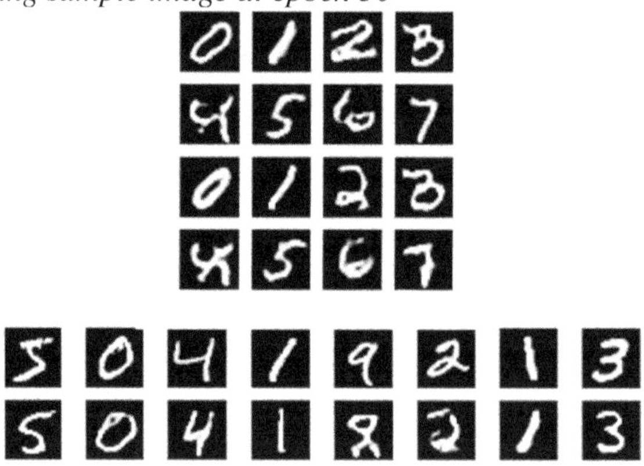

Figure 15. Training sample image at epoch 100

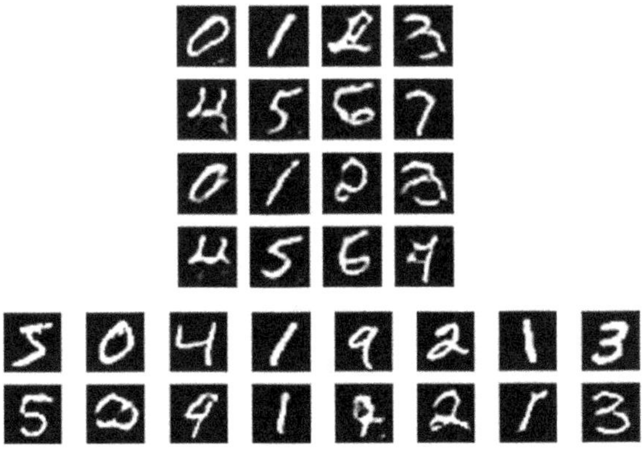

4.3 Comparative Analysis

To further test the performance of the GAN model, comparative analysis was carried out on real MNIST images against the synthetic images of the model. The comparison picture generation function permitted the production of various visual comparisons between real and produced images to perform side-by-side comparisons. The resemblance of these created images to the real ones is brought out both in their holistic structure and fine details. Besides the subjectively visual comparison, a measurable comparison was done by computing the Fréchet Inception Distance (FID) score when comparing the distribution of the generated images against their matching real ones. As a number of training epochs passed, the FID score continued to decline, implying that the distribution of produced images gradually moved closer to the distribution of real images. ***Figure 16*** shows generated output of GAN. It clearly displays the output formation mechanism from the input provided by users through the model. This figure indicates that, once trained, the network can synthesize realistic data, displaying effectiveness and quality in the generated outputs.

Figure 16. GAN's final result and evaluation, exhibiting how the model creates outputs based on user input, showcasing the network's capacity to synthesize realistic data after training.

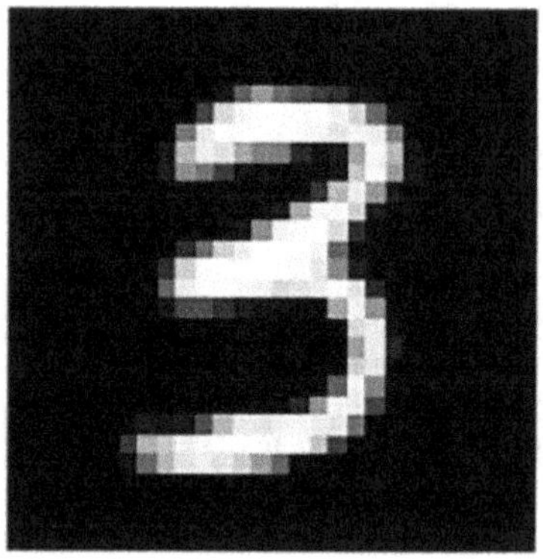

```
Please enter a digit (0-9):  5
```

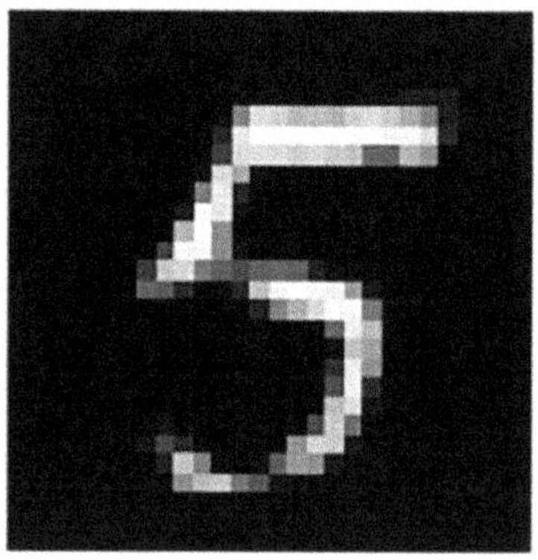

CONCLUSION

The chapter has gone step-by-step on how to design and train a GAN model from scratch using TensorFlow. You followed the guide to do data pre-processing, design generator and discriminator networks, establish loss functions, and then put it all together into the training process. Furthermore, you have seen how to check training progress, store models, and produce fresh images according to user input. This will provide the basic knowledge that will allow you to continue to experiment with GANs and explore their possibilities in diverse applications.

REFERENCES

Ali, A., Senan, N., & Murli, N. (2024). *Convolutional Neural Network Using Regularized Conditional Entropy Loss (CNNRCoE) for MNIST Handwritten Digits Classification* (pp. 337–348). DOI: 10.1007/978-3-031-66965-1_33

Chakraborty, T., Reddy, K. S. U., Naik, S. M., Panja, M., & Manvitha, B. (2024). Ten years of generative adversarial nets (GANs): A survey of the state-of-the-art. *Machine Learning: Science and Technology*, 5(1), 011001. DOI: 10.1088/2632-2153/ad1f77

Dash, A., Ye, J., & Wang, G. (2024). A Review of Generative Adversarial Networks (GANs) and Its Applications in a Wide Variety of Disciplines: From Medical to Remote Sensing. *IEEE Access : Practical Innovations, Open Solutions*, 12, 18330–18357. DOI: 10.1109/ACCESS.2023.3346273

Figueira, A., & Vaz, B. (2022). Survey on Synthetic Data Generation, Evaluation Methods and GANs. *Mathematics*, 10(15), 2733. DOI: 10.3390/math10152733

Goodfellow, I., Pouget-Abadie, J., Mirza, M., Xu, B., Warde-Farley, D., Ozair, S., Courville, A., & Bengio, Y. (2020). Generative adversarial networks. *Communications of the ACM*, 63(11), 139–144. DOI: 10.1145/3422622

Goodfellow, I. J., Pouget-Abadie, J., Mirza, M., Xu, B., Warde-Farley, D., Ozair, S., Courville, A., & Bengio, Y. (2014a). *Generative Adversarial Networks*. http://arxiv.org/abs/1406.2661

Goodfellow, I. J., Pouget-Abadie, J., Mirza, M., Xu, B., Warde-Farley, D., Ozair, S., Courville, A., & Bengio, Y. (2014b). Generative Adversarial Networks. *Science Robotics*, 3(January), 2672–2680. https://arxiv.org/abs/1406.2661v1

Huang, G., Liu, Z., Van Der Maaten, L., & Weinberger, K. Q. (2017). Densely connected convolutional networks. *Proceedings - 30th IEEE Conference on Computer Vision and Pattern Recognition, CVPR 2017, 2017-January*, 2261–2269. DOI: 10.1109/CVPR.2017.243

Karras, T., Aittala, M., Hellsten, J., Laine, S., Lehtinen, J., & Aila, T. (n.d.). *Training Generative Adversarial Networks with Limited Data*. https://github.com/NVlabs/stylegan2-ada

König, T., Wagner, F., Bäßler, R., Kley, M., & Liebschner, M. (2023). Synthetic data generation of vibration signals at different speed and load conditions of transmissions utilizing generative adversarial networks. *TM. Technisches Messen*, 90(10), 639–649. DOI: 10.1515/teme-2023-0001

Lecun, Y., Bottou, L., Bengio, Y., & Haffner, P. (1998). Gradient-based learning applied to document recognition. *Proceedings of the IEEE*, 86(11), 2278–2324. DOI: 10.1109/5.726791

Li, Z.-D., He, W.-Y., & Ren, W.-X. (2024). Structural damage identification based on Wasserstein Generative Adversarial Network with gradient penalty and dynamic adversarial adaptation network. *Mechanical Systems and Signal Processing*, 221, 111754. DOI: 10.1016/j.ymssp.2024.111754

Luedtke, A., Carone, M., Simon, N., & Sofrygin, O. (2020). Learning to learn from data: Using deep adversarial learning to construct optimal statistical procedures. *Science Advances*, 6(9), eaaw2140. Advance online publication. DOI: 10.1126/sciadv.aaw2140 PMID: 32166115

O'Shea, K., & Nash, R. (2015). An Introduction to Convolutional Neural Networks. *International Journal for Research in Applied Science and Engineering Technology*, 10(12), 943–947. DOI: 10.22214/ijraset.2022.47789

Reed, S., Akata, Z., Yan, X., Logeswaran, L., Schiele, B., & Lee, H. (2016). Generative Adversarial Text to Image Synthesis. *33rd International Conference on Machine Learning, ICML 2016, 3*, 1681–1690. https://arxiv.org/abs/1605.05396v2

Sengupta, S., Basak, S., Saikia, P., Paul, S., Tsalavoutis, V., Atiah, F., Ravi, V., & Peters, A. (2020). A review of deep learning with special emphasis on architectures, applications and recent trends. *Knowledge-Based Systems*, 194, 105596. DOI: 10.1016/j.knosys.2020.105596

Torfi, A., Fox, E. A., & Reddy, C. K. (2022). Differentially private synthetic medical data generation using convolutional GANs. *Information Sciences*, 586, 485–500. DOI: 10.1016/j.ins.2021.12.018

Training a neural network on MNIST with Keras | TensorFlow Datasets. (n.d.). Retrieved January 27, 2025, from https://www.tensorflow.org/datasets/keras_example

Yuan, L., Yu, S., Yang, Z., Duan, M., & Li, K. (2023). A data balancing approach based on generative adversarial network. *Future Generation Computer Systems*, 141, 768–776. DOI: 10.1016/j.future.2022.12.024

Zhou, T., Li, Q., Lu, H., Cheng, Q., & Zhang, X. (2023). GAN review: Models and medical image fusion applications. *Information Fusion*, 91, 134–148. DOI: 10.1016/j.inffus.2022.10.017

Zühlke, M.-M., & Kudenko, D. (2024). Adversarial Robustness of Neural Networks From the Perspective of Lipschitz Calculus: A Survey. *ACM Computing Surveys*. Advance online publication. DOI: 10.1145/3648351

Chapter 3
A Novel Photorealism Framework for Image Generation Using Generative Adversarial Networks

Soumya Ranjan Mishra
KIIT University, Bhubaneswar, India

Aaryan Sinha
https://orcid.org/0009-0000-1403-5704
KIIT University, Bhubaneswar, India

Samik Dey
https://orcid.org/0009-0005-8394-0448
KIIT University, Bhubaneswar, India

Hitesh Mohapatra
https://orcid.org/0000-0001-8100-4860
KIIT University, Bhubaneswar, India

Mahendra Kumar Gourisaria
https://orcid.org/0000-0002-1785-8586
KIIT University, Bhubaneswar, India

ABSTRACT

This paper presents a framework for generating realistic photographs using Generative Adversarial Networks (GANs). Our framework comprises two key neural networks:

DOI: 10.4018/979-8-3693-7152-7.ch003

the Generator and the discriminator. The generator network creates images, while the discriminator network evaluates their authenticity. This framework highlights the potential and advancements of GANs in creating photorealistic visuals. To demonstrate the comparative analysis, we created two distinct models with different picture sizes 128 pixels and 64 pixels, respectively using two different GPUs, the T4x2 and P100. The produced pictures from the P100 GPU have been surpassed by those from the T4x2 GPU, the P100 model has been outperformed by the T4x2 GPU, by considering metrics like LOSS_G, LOSS_D, Real Score, and Fake Score. From the P100 model to the 'T4x2 model, the average value of phony SCORE shows a noteworthy improvement of about 59.34\%, suggesting improved performance in identifying phony data in the later model.

INTRODUCTION

Creation of Realistic images has been a challenge in the domain of artificial intelligence and computer graphics, holding a vast promise in gaming, virtual reality and digital entertainment (He Huang & Wang, 2018). The methods used for capturing the fine details of the real facial features based on traditional ways crafted are more artificial looking results. Due to new advancement in deep learning particularly in the field of the advent of Generative Adversarial Network which have revolutionized the creation of photorealistic images. Since their introduction of GANs have proven highly effective in producing high-fidelity images that closely resemble real-world data. By putting the two neural networks together, the generator and discriminator, together with a competitive nature, can learn the complex data distribution to generate the real images that can't be distinguished from real samples. The paper describes the complex process of utilizing the GANs for realistic face generation, exploring their architecture and various techniques for enhancement of the generated images. Challenges like identity retention, expression and fine grained details are analyzed from a concerned perspective. With the growing demand for the authenticity of virtual experiences and personalized content creation, the ability to generate realistic faces holds a significant importance. By straightening out the methodologies and advancement in face generation the study contributes to report in computer graphics and artificial intelligence, covering the way for the future innovations in visual content generations and human-computer interaction Since GANs have made strides in generating the photogenic images of human faces and many other image generations, many challenges are unsolved. Some challenges include identity upholding, expression variable from person to person and the main thing fine grained details is one of the ongoing difficulties which is ongoing. The main aim of this paper is to showcase the involved process of utilizing GANs for face generation. It will explore imaginative methods for identity boosting and managing facial expression variables and fine grain refining in generated faces.

LITERATURE SURVEY

A high powered architecture for generating the realistic images, including the faces is Generative Adversarial Networks (GANs) (Silver et al., n.d.). Our Literature review aims to investigate the challenges and progress in crafting of realistic images with GANs which focuses on the different architectures, training strategies and evaluation methods employed in current studies. The revolution of the model of the generative model after the enhancement of the GAN designs to generate high-resolution and photorealistic images. The earliest significant advancement in face generations with GANs was made using the Deep Convolutional GAN (DCGAN) architecture. The DCGAN exhibits the capability of GANs to produce provident faces with improvement in stability and quality. Several improvements to GAN designs, especially designed for face crafting have been made via more research advancements. Progressive Growing GAN (PGGAN), creates the photo in coarse-to-fine nature. It produces the sharper and more real faces by repeatedly raising the resolution during training.

A powerful framework for generating realistic images, including faces, is Generative Adversarial Networks (GANs). This literature review aims to investigate the progress and challenges in creating realistic faces with GANs, focusing on different architectures, training strategies, and evaluation methods employed in current studies. The field of generative modelling underwent a revolution with the introduction of GANs. Since then, researchers have endeavoured to enhance GAN designs to generate high-resolution and photorealistic images. One of the earliest significant advancements in face generation with GANs was made using the Deep Convolutional GAN (DCGAN) architecture. DCGAN demonstrated the capability of GANs to produce recognizable faces with improved stability and quality. Several enhancements to GAN designs especially designed for face creation have been made via subsequent research initiatives. To create high-resolution faces with fine features, progressive Growing GAN (PGGAN), creates pictures in a coarse-to-fine way which produces sharper and more realistic faces by progressively raising the resolution during training, which solves the problem of training high-resolution GANs. By straightening out the latent space representation, StyleGAN also introduced. it improved the quality and diversity of produced faces even further. StyleGAN enables more precise control over some produced face traits, such as age, gender, and facial expression, resulting in the production of incredibly realistic and varied face pictures. Manipulating specific face characteristics is made possible by the disentangled representation, which maintains overall realism. Even with the great advancements, there are still several difficulties in employing GANs to generate realistic faces. Realistically attaining diversity and realism at the same time is one of the main issues. Even while GANs may create aesthetically pleasing faces, a key area of research continues to be ensuring variety in produced samples without sacrificing realism. To handle the situation, methods like conditional generation, diversity and quality of predicted faces is essential to determining how well GAN models function. While it offers quantitative assessments of a picture like Frechet Inception Distance (FID) and Inception Score (IS) can fall short for the intuitive. Intuitive based metrics and studies have presented recently to assess the identity preservation and facial features. Face-generation research with GANs, ethical issues are also quite important. Issues like mitigation of biases and varied representation are actively investigated. In summary the GANs have transformed the production of the realistic faces by referring to the unseen potential for producing the variable facial representation. Since problems like realism and ethics issues are there which still exist which continued for research into the more reliable face crafting with GANs. With the rise in the use of Generative Adversarial Network, there are many frameworks and analyses were done (He Huang & Wang, 2018), offered a taxonomy of image synthesis techniques and reviewed various models for image and text-to-picture synthesis,image- to-image conversion (Silver et al., n.d.). It has also unveiled a novel search technique that integrates value and policy networks with Monte Carlo simulation (Goodfellow, Pouget-Abadie, & Bengio, 2014) presented a novel framework for generative model estimation using adversarial nets. Experiments show the framework's potential by evaluating the produced samples both statistically and qualitatively (Radford & Chintala, 2015) aimed to close the achievement gap between CNNs for supervised learning and unsupervised learning with their study. They have also applied the acquired characteristics to new jobs, showing that they are applicable as universal picture representations (Salimans, Goodfellow, & Chen, 2016). Their methodology produces samples for MNIST that are indistinguishable by humans from actual data, and samples for CIFAR-10 that result in a human error rate of 21.3% (Yu, Zhang, & Yu, 2017). To address the issues, they present a sequence-generating framework called SeqGAN. SeqGAN avoids the generator differentiation problem by executing gradient policy updating directly, by modelling the data generator as a stochastic policy in reinforcement learning (RL) (Liang, Hu, & Xing, 2017). A structured paragraph generator which generates the sentences recurrently by incorporating the region based language. (Guo, Lu, & Wang, 2017). They discover that DeepNAG performs better than DeepGAN while training five recognizers using data augmentation over six datasets. (Upadhyay, 2023) They provide techniques for producing high-quality photographs using generative adversarial networks.

One kind of Generative Adversarial Network (GAN) utilized for picture creation is called DCGAN (Deep Convolutional Generative Adversarial Network) (Singh & Raza, n.d.), their work offers cutting-edge advancements in cross-modality synthesis and GANs-based clinical applications for the creation of medical images (Antony Arul Raj & Rukmani, 2024). They begin with its foundations and cover potential and challenges (Mercaldo & Martinelli, 2023), also describe a technique that uses a dataset of retinal pictures to evaluate the distinguishability of bioimages produced by generative adversarial networks (Liu & Huang, 2020). An overview of GANs is given in this study, with a particular emphasis on techniques and applications related to visual synthesis (Ghosh, Kulharia, & Dokania, 2018). Researchers conduct in-depth tests on both artificial and real datasets, contrasting MAD-GAN with other GAN variations. Also demonstrated high-quality diversified sample generation for difficult tasks like face creation and image-to-image translation (Galteri, Seidenari, & Bimbo, 2017), they demonstrate that, as a baseline, their model may also be trained by maximizing the Structural Similarity (SSIM), which is a superior loss compared to the more straightforward Mean Squared Error (MSE) (Karras, Laine, & Aila, 2020), also shows us how well the generator uses its output resolution and points up a capacity issue, which inspires us to train bigger models for further quality enhancements (Kim & Lee, 2016). Presented a highly accurate single image super resolution. (Kim, Cha, & Kim, 2017), suggested a technique called DiscoGAN, which learns to discover the relation between different domains. (Kingma & Dhariwal, 2018), also suggested Glow, a log-likelihood objective enough to create realistic and manipulated images. (Nazeri, Ng, Joseph, Qureshi, & Ebrahimi, n.d.) created a novel method for picture inpainting that is based on the region normalization. Text to Photo-Realistic Image Synthesis using Generative Adversarial Networks (Rao et al., 2023), This study explores the use of GANs to generate high-resolution images from textual descriptions, reducing reliance on manual design and fostering creativity. The ability of GANs to generate highly realistic human faces raises several ethical concerns. One of the potential misuse of GAN is generation of content for malicious purposes like creation of deep fakes, which can spread harmful reputations and affect the privacy.Also GANs can be used to bypass content in generation of realistic images of individuals leading to legal concerns around the data ownership. The lack of the transparency in the models also raises a question around the authenticity and trust especially in the sensitive domains like security, virtual identity and media. Dealing with these concerns requires an implementation of robust safeguards and watermarking can be applied, for ethical AI use cases (Mishra, Mohapatra, Edalatpanah, & Gourisaria, 2024).

PROPOSED MODEL

In the field of AI, GANs have outraged the methods especially in the field of computer vision (He Huang & Wang, 2018). In the model we have two neural networks known as Generator and Discriminator which interact dynamically from the base of the GANs. Here the generator acts as an artist whose role is to create a realistic job out of the random noise which is sampled from a latent space. The conversion of the latent space into high dimensional representations that act like real images which are generally done by the Transposed convolutional layers which are part of the model. On the other hand the discriminator acts as a critic and determines if the image created is real or fake. The capacity of the discriminator increases for the real and fake as both the neural networks are trained where the generator 's goal is to create images that are more realistic and to fake the discriminator. Figure 1, shows the Block diagram of the flowchart of the Generative Adversarial Network which we have applied to our framework. Here, we are using two different layers for the generator and discriminator shown in Figure 2 (Convolution 2D model) and Figure 3 (Convolution Transpose 2D model). The complete flow of the architecture is illustrated in the following figures. Figure 4 depicts the detailed architecture of the Generator, while Figure 5 illustrates the comprehensive architecture of the Discriminator.

The framework designed by us has two different environments where we have done training on the model. The very first is the environment where we refer to it as 128 GANs Model with an accelerator used which is known as T4x2 a very famous GPU present in Kaggle. Referring to this image we have considered the image size to be 128 which could help us to get more accurate results with detailing in images. T4x2 is a leading GPUs of NVIDIA which is used for Image Processing. The capacity of the T4 GPU is doubled by the T4x2 GPU which can perform very well in image processing but requires dealing with huge datasets only. The T4x2 expedites processes such as picture filtering, scaling, and colour correction by mitigating bottlenecks stemming from memory limits. The T4x2 is an invaluable tool for image processing jobs that make use of deep learning techniques, such as image identification and picture modification. This is because its Tensor Cores offer hardware-level acceleration for deep learning frameworks.

The other environment that we have used is called 64 GANs MODEL with the accelerator known as P100 (GPU). In our environment the image size that we have declared is 64 so that we can make a comparison between the two GPUs. The P100 GPUs is powerful when it comes to image processing as the pascal architecture and 16GB of RAM which makes it good for deep learning research and model creation. It also performs exceptionally well in well-established processes, handling colour correction, scaling, and filtering with ease. Given its affordable pricing and the lack of need for the highest speeds for our work.

Figure 1. Block diagram of generative adversarial networks

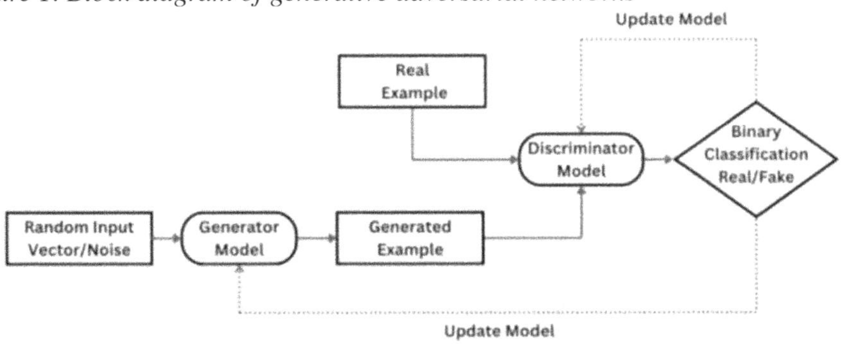

Figure 2. Generator architecture 1

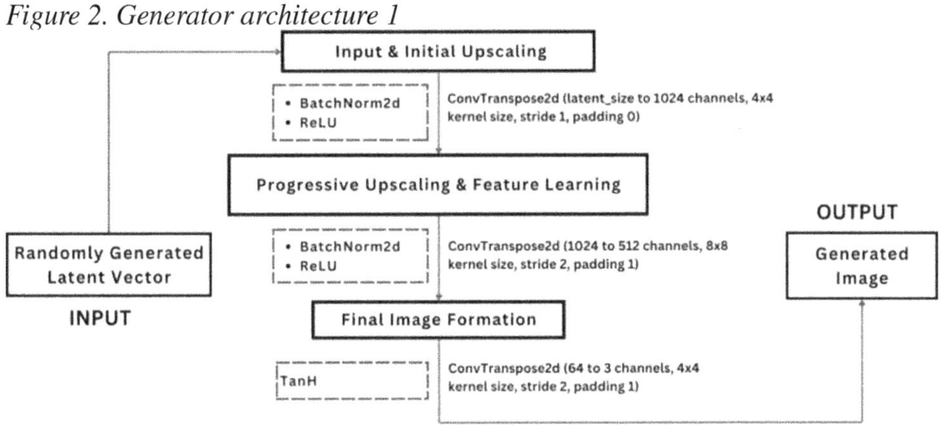

Generative Adversarial Networks (GANs) are a key technique in the field of artificial intelligence and computer vision today, earlier known with the powers of producing real images. The demand for innovative solutions with the help of AI and the generosity of digital material have made automation of picture generation systems essential. It has taken orderly steps to approach the capability of GANs. Data preparation is the first step in the process of creating GANs, after the planning of the collection ensures a clean and structured sample of real world data by removing the oddities This step is done to improve the diversity by standardizing pixel values and increasing the dataset. The architecture of the model consists of the two models that are neural networks commonly known as Generator and Discriminator where the models are locked in a dynamic interaction in which the activation dataset has been carefully selected. As generator is referred to as artist which creates a realistic images using the sample random noise from a latent, and the discriminator simultaneously serves as a critic which distinguishes the artificial crafted image and the real images. The discriminator loss assesses how accurate the discriminator was in the

Figure 3. Discriminator architecture 1

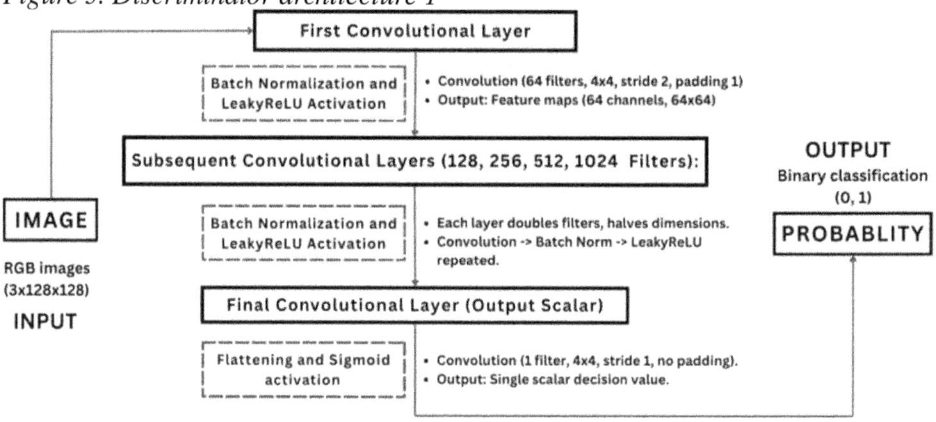

Figure 4. Generator architecture 2

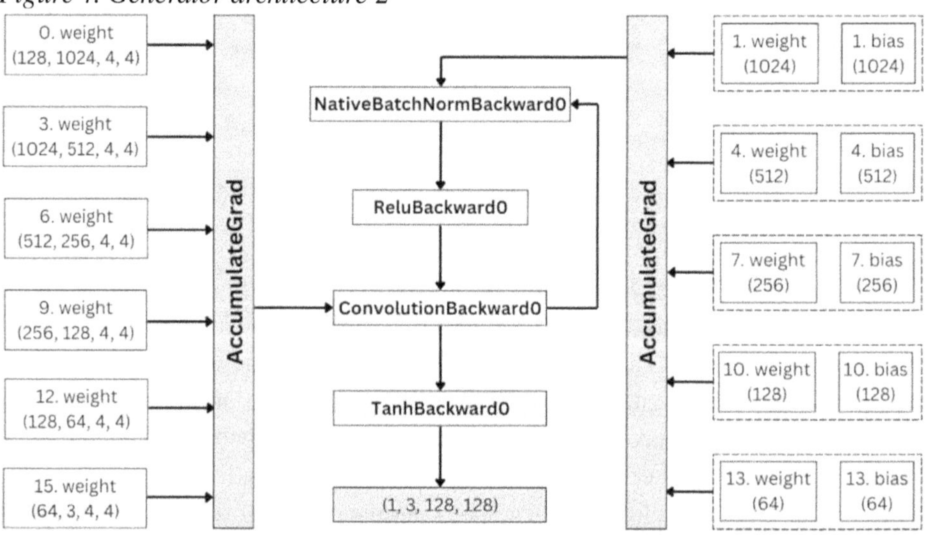

classification of the generated pictures, whereas the generator loss is analyzed and the generated photos against the discriminator classification. In qualitative assessment it involves visual inspection of the produced images which can be assessed to their realism and faithfulness to the dataset, and quantitative metrics such as Inception Score and Frechet Inception Distance, offers an objective of measurement of image quality and diversity. GANs are enormous performing in fields such as computer graphics, creative design and other which has led the creativity to the next level.

Figure 5. Discriminator architecture 2

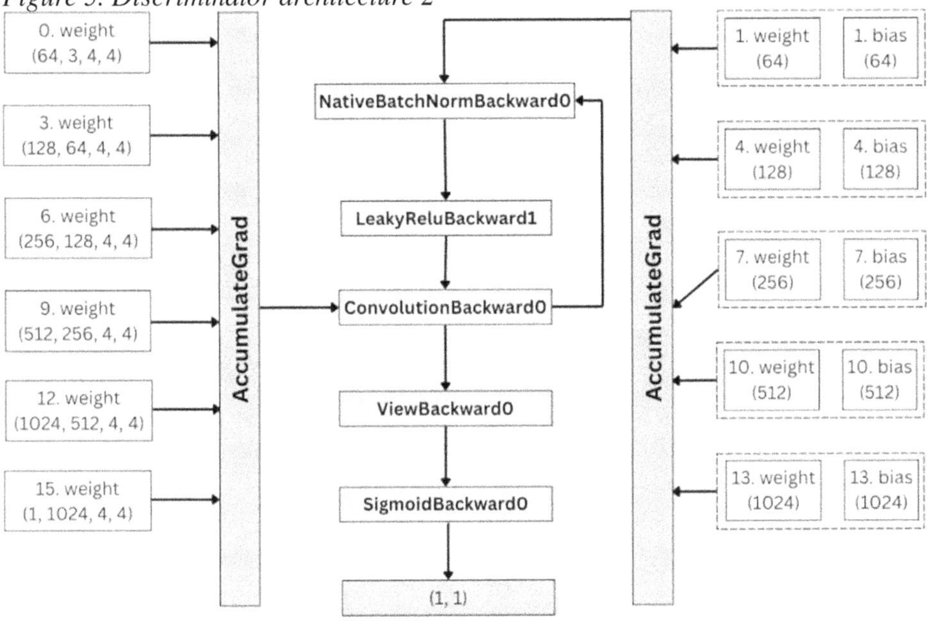

RESULTS ANALYSIS

(a) Output Image for Epoch 2 (b) Output Image for Epoch 4
(c) Output Image for Epoch 6 (d) Output Image for Epoch 8
(e) Output Image for Epoch 10 (f) Output Image for Epoch 12

Figure 6. Output images for different epoch numbers

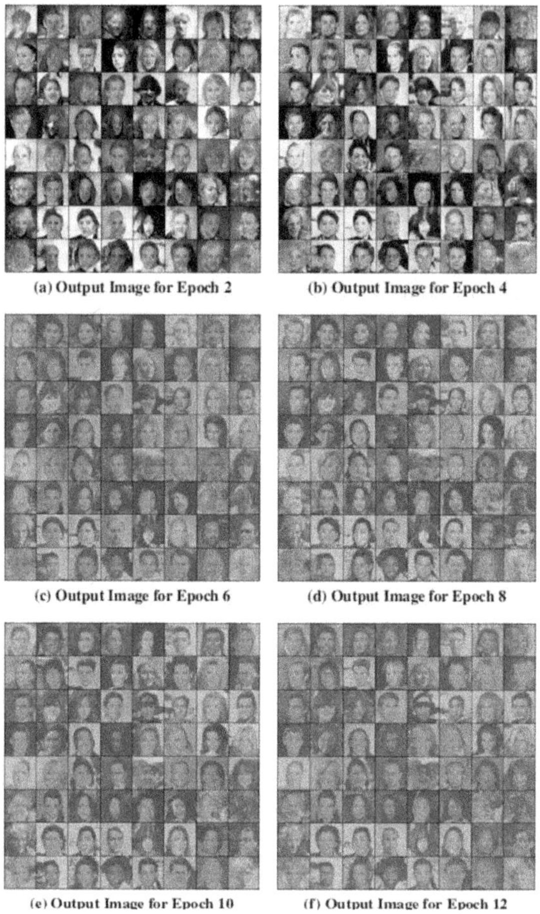

(a) Output Image for Epoch 2 (b) Output Image for Epoch 4

(c) Output Image for Epoch 6 (d) Output Image for Epoch 8

(e) Output Image for Epoch 10 (f) Output Image for Epoch 12

We thoroughly assessed the algorithms in our GAN framework using four important metrics: Loss G, Loss D, Real Score, and Fake Score. The values for them are shown in the table These metrics offer important insights into the dynamics and operation of the GAN model's generating and discriminative parts during training. LossG is used to calculate the performance of the generator which is the difference between the produced and actual data distribution. In the same way the LossD is used to quantify the dicriminator's capacity to distinguish between the real and fraudulent samples. Real and Fake Score are used for identifying the confidence of the discriminator actual and produced data respectively. For the same, we have included the table for two different models. We have collected informative metrics values that are essential for evaluating algorithm performance in our study of the GAN framework. The generator's performance is measured by the Loss G metric, which quantifies the difference between the produced and actual data distributions. In the same way the LossD measures the discriminator performance for the skill to distinguish the authenticity and fraudulent samples. The accuracy of the discriminator is referred to as Fake Score, whereas the Real Score indicates the confidence in correctly identifying real data instances. When combined the metrics details a picture of dynamics and efficiency of model, and the adjustment of model can also be done. To show the evolution of metric scores of two different models, they are obtained by comparing the two pixel image sizes that are 128x128 and 64x64 GANs models. Hence we can say that the model could be more difficult for the large one and might require more iterations. Even while both models get similar average scores in successfully identifying genuine photographs, the 128x128 GAN model tends to have greater confidence in its assessments, as seen by its higher maximum actual score in terms of discriminator performance. The 128x128 GAN model, on the other hand, has a larger average fake

(a) Discriminator Real Fake Score for 128 GANs Model (b) Discriminator Real Fake Score for 64 GANs Model
(c) Training Loss Score for 128 GANs Model (d) Training Loss Score for 64 GANs Model

Figure 7. Comparison of discriminator real fake scores and training loss for different GAN models

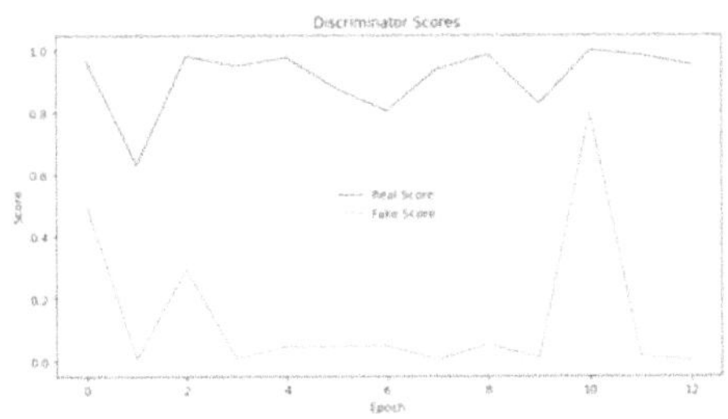

(a) Discriminator Real Fake Score for 128 GANs Model

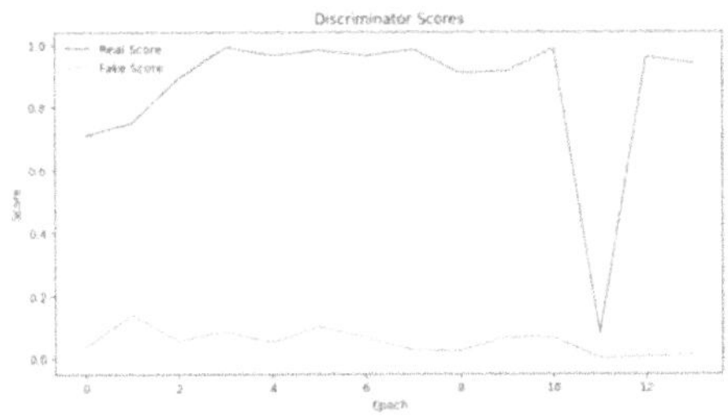

(b) Discriminator Real Fake Score for 64 GANs Model

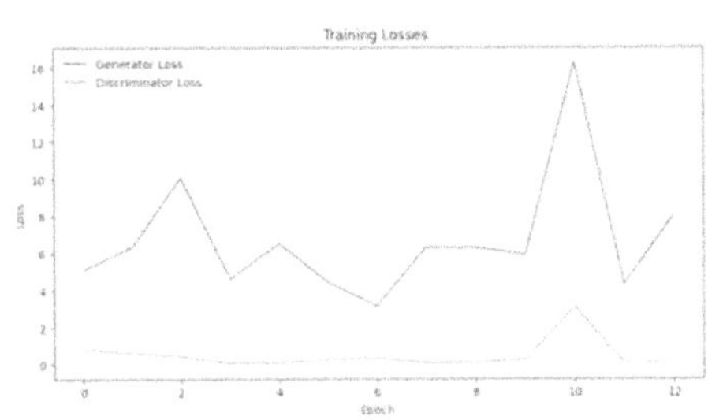

(c) Training Loss Score for 128 GANs Model

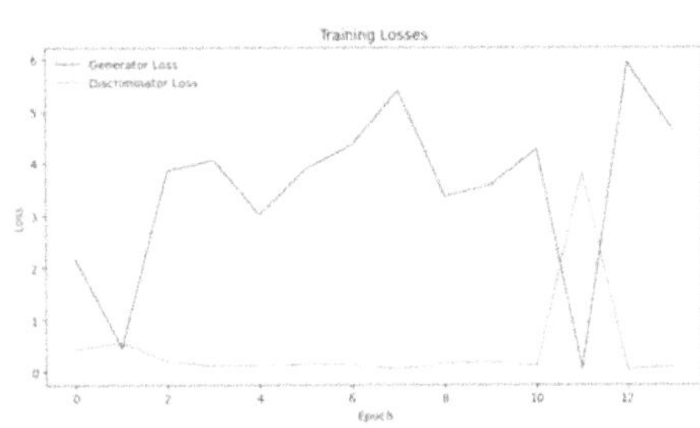

(d) Training Loss Score for 64 GANs Model

score compared to the 64x64 architecture, which is an indicating factor that it is very much less certain of its holding to identify the false images. Although if we see the occasional higher maximum fake score, the consolidating has a higher level of confirmation in identifying the fake/bogus images. These factors suggest that characteristics like picture size and complexity affects the training dynamics and performance of the models. Although if we see the larger models must require fine tuning and oversight to achieve an optimal level of performance, smaller models may exist with more consistent training behaviour. The output for Epoch 2, Epoch 4, Epoch 6, Epoch 8, Epoch 10 and Epoch 12 are shown in Figure 6 a,b,c,d,e and f. During training, the GAN occasionally encountered mode collapse, where the generator repeatedly produced a limited variety of outputs, leading to a lack of diversity in generated images. Feature matching loss is Introduced to ensure the generator captured a broader distribution of data instead of focusing on fooling the discriminator.

Table 1. Metric score for 128 GANs model

Metric Name	Average Score	Maximum Score	Minimum Score
Loss Generator	6.7922	16.2838	3.1502
Loss Discriminator	0.4134	3.1203	0.0338
Real Score	0.9002	1.0000	0.6304
Fake Score	0.1198	0.8008	0.0006

Table 2. Metric score for 64 GANs model

Metric Name	Average Score	Maximum Score	Minimum Score
Loss Generator	3.9366	5.9334	0.0543
Loss Discriminator	0.3817	3.8290	0.0442
Real Score	0.9026	0.9896	0.0776
Fake Score	0.0487	0.1380	0.0004

CONCLUSION AND FUTURE SCOPE

Training Loss The learning behaviour of the two models (128 and 64) may be seen by examining their respective training loss graphs shown in Figure 7. The ideal loss curves for each model should show a general declining trend over time.

In comparison to the 64 models, a greater drop in the 128 model's loss curve can suggest quicker learning (Silver et al., n.d.). It is typical for both loss curves to exhibit oscillations (up and down trends) during training. The same can be observed in Table.1. The discriminatory score graphs would demonstrate the models' capacity to discern between genuine and false pictures. These would probably be for the categorization of real vs. fake images for both the 128 and 64 models. The holding value of discrimination for the 128 model is greater and would be shown by the very large difference between the true and false value on the 129 models graph compared to 64 models shown in Table.2. Therefore we can say that 128 models may be more adaptable to the detection of phony photos. Hence concluding the project on training is purely based on Generative Adversarial Network which significantly depends on our understanding of the different set of images and their detailing, by generating high quality synthetic images. By training both the neural networks that are generator and discriminator at the same time which made us understand how they are outperforming each other. Through this adversarial process, the generator learned to produce increasingly realistic images, while the discriminator honed its ability to distinguish between real and generated content. An intriguing picture of their capabilities is painted by analyzing the training loss and discriminating score graphs, as well as the fact that the 128 model was trained on a T4x2 GPU with double the RAM of the 64 model's GPU (which may have been a separate model). The loss curve indicates that the performance of the 128 model may be exponentially influenced by T4x2's memory advantage. Quick decrease in the loss curve indicates faster learning, enabling it to investigate a larger variety of picture variants in the training process. This results in the creation of more complex and true to life pictures. The 128 model may have initially fooled the discriminator with a wider gap between real and fake scores, as evidenced by the discriminatory score graphs. However, as the model became more adept at producing realistic fakes, the discriminator may have had to adjust and enhance its own capabilities. The difference in both models may eventually close or even reverse as the discriminator gains the ability to identify the patterns created by the 128 models. Due to T4x2 having more memory the discriminator is tested for a longer amount of time so the 128 model may have an early advantage in producing more convincing fakes. To crown the better model, all the supplied graphs in addition to other metrics like accuracy, precision, and recall must be considered. Though GANs have demonstrated phenomenal ability to generate realistic images, there is still scope for improvement. Further research must be focused on developing more optimised training solutions and efficient architectures to generate better quality and detailed images.

REFERENCES

Anand, S. V. (2023). *Upadhyay1, "Image generation using generative adversarial network.* IJIRT.

Creswell, A., White, T., Dumoulin, V., Arulkumaran, K., Sengupta, B., & Bharath, A. A. (2018). Generative adversarial networks: An overview. *IEEE Signal Processing Magazine*, *35*(1), 53–65.

Galteri, L., Seidenari, L., Bertini, M., & Del Bimbo, A. (2017). Deep generative adversarial compression artifact removal. In *Proceedings of the IEEE international conference on computer vision* (pp. 4826-4835).

Ghosh, A., Kulharia, V., Namboodiri, V. P., Torr, P. H., & Dokania, P. K. (2018). Multi-agent diverse generative adversarial networks. In *Proceedings of the IEEE conference on computer vision and pattern recognition* (pp. 8513-8521).

Goodfellow, I., Pouget-Abadie, J., Mirza, M., Xu, B., Warde-Farley, D., Ozair, S., . . . Bengio, Y. (2014). Generative adversarial nets Advances in neural information processing systems. arXiv preprint arXiv:1406.2661.

Guo, J., Lu, S., Cai, H., Zhang, W., Yu, Y., & Wang, J. (2018, April). Long text generation via adversarial training with leaked information. *Proceedings of the AAAI Conference on Artificial Intelligence*, *32*(1), •••.

Huang, H., Yu, P. S., & Wang, C. (2018). An introduction to image synthesis with generative adversarial nets. arXiv preprint arXiv:1803.04469.

Karras, M. A. J. H. J. L. T., Laine, S., & Aila, T. (2020). *Analyzing and improving the image quality of stylegan.* CVPR. DOI: 10.1109/CVPR42600.2020.00813

Kim, H. K. J. K. L. T., Cha, M., & Kim, J. (2017). *Learning to discover cross-domain relations with generative adversarial networks.* ICML.

Kim, J. K. L. J., & Lee, K. M. (2016). *Accurate image superresolution using very deep convolutional networks.* CVPR.

Kingma, D. P., & Dhariwal, P. (2018). Glow: Generative flow with invertible 1x1 convolutions. *Advances in Neural Information Processing Systems*, •••, 31.

Liang, X., Hu, Z., Zhang, H., Gan, C., & Xing, E. P. (2017). Recurrent topic-transition gan for visual paragraph generation. In *Proceedings of the IEEE international conference on computer vision* (pp. 3362-3371).

Liu, M. Y., Huang, X., Yu, J., Wang, T. C., & Mallya, A. (2021). Generative adversarial networks for image and video synthesis: Algorithms and applications. *Proceedings of the IEEE, 109*(5), 839–862.

Mercaldo, F., Brunese, L., Martinelli, F., Santone, A., & Cesarelli, M. (2023). Generative Adversarial Networks in Retinal Image Classification. *Applied Sciences (Basel, Switzerland), 13*(18), 10433.

Mishra, S.R., Mohapatra, H., Edalatpanah, S.A. i Gourisaria, M.K. (2024). Advanced deepfake detection leveraging swin transformer technology. Engineering Review, 44 (4 - SI 2024), 45-56. DOI: 10.30765/er.2583

Nazeri, K., Ng, E., Joseph, T., Qureshi, F. Z., & Ebrahimi, M. (2019). Edgeconnect: Generative image inpainting with adversarial edge learning. arXiv preprint arXiv:1901.00212.

. N.K.SINGH. and K.Raza, "Medical image generation using generative adversarial networks,"

. L. M. A. Radford and S. Chintala, "Unsupervised representation learning with deep convolutional generative adversarial networks," 2015.

. A. S. Rao, P. A. Bhandarkar, P. A. Devanand, P. Shankar, S. Shanti and K. P. B H, "Text to Photo-Realistic Image Synthesis using Generative Adversarial Networks," 2023 2nd International Conference on Futuristic Technologies (INCOFT), Belagavi, Karnataka, India, 2023, pp. 1-6, .DOI: 10.1109/INCOFT60753.2023.10425482

Salimans, W. Z. V. C. A. R. T., Goodfellow, I., & Chen, X. "Improved techniques for training gans," in advances in neural information processing systems," p. 2226–2234, 2016.

. D. Silver, A. Huang, C. J. Maddison, A. Guez, L. Sifre, G. V. D. Driessche, J. Schrittwieser, I. Antonoglou, V. Panneershelvam, M. Lanctot, and et al., "Title of the article,"

Yu, J. W. L., Zhang, W., & Yu, Y. "Seqgan: Sequence generative adversarial nets with policy gradient," p. 2852–2858, 2017.

Chapter 4
Crafting Images With Generative Adversarial Networks (GANs) and Models

Naveen Kumar Thawait
https://orcid.org/0009-0001-7761-3097
Dr. C.V. Raman University, India

ABSTRACT

The chapter "Challenges and Future Directions" in the book "Crafting Images with Generative Adversarial Networks (GANs) and Models" delves into the current obstacles and prospective advancements in the realm of GANs and generative models. This chapter aims to provide readers with a thorough understanding of the primary issues hindering the progress and broader application of GANs, while also highlighting the innovative solutions and future directions that could address these challenges. The first section of the chapter explores the major challenges faced by researchers and practitioners in the field of GANs. These include training instability, where GANs often suffer from issues like mode collapse and vanishing gradients, making the training process unpredictable and inefficient.

INTRODUCTION

The introduction of this chapter delves into the dual nature of GANs, highlighting their transformative impact across domains such as healthcare, art, and education, while also acknowledging the technical hurdles and societal implications they pose. It sets the stage by framing GANs not only as a groundbreaking technological ad-

DOI: 10.4018/979-8-3693-7152-7.ch004

vancement but also as a field requiring careful navigation to alleviate challenges like training instability, mode collapse, and ethical concerns. The thesis statement emphasizes the chapter's focus on exploring these challenges and presenting innovative solutions to ensure GANs are leveraged responsibly and effectively for societal benefit. This chapter examines the dual facets of Generative Adversarial Networks (GANs), spotlighting their transformative role in reshaping fields like healthcare, art, and education, while addressing the significant challenges they present. GANs have revolutionized content creation, enabling advancements in areas like medical imaging, synthetic data generation, and creative design. However, alongside their promise lies a host of technical and societal issues that demands careful attention. Challenges such as training instability, mode collapse, and the computational demands of these models hinder their efficient development and deployment. At the same time, societal concerns, including the potential misuse of GANs in creating deepfakes or violating intellectual property rights, raise ethical questions about their responsible use. The thesis statement of this chapter underscores its comprehensive approach: "This chapter explores the challenges and advancements of GANs, emphasizing their transformative potential while addressing technical and societal challenges. It offers solutions to ensure their responsible and innovative use in fields like healthcare, art, and education." By integrating discussions on technical solutions and ethical frameworks, the chapter aims to provide a balanced perspective on maximizing GANs' benefits while mitigating associated risks.

Generative Adversarial Networks (GANs) have revolutionized the field of artificial intelligence by enabling the creation of highly realistic images, sounds, and other data forms. Introduced by Ian Goodfellow and his colleagues in 2014, GANs consist of two neural networks—the generator and the discriminator—that are trained in tandem to produce data indistinguishable from real-world examples. As these networks compete against each other, they progressively enhance the quality of the generated outputs, pushing the boundaries of what machines can create autonomously. GANs have become a cornerstone of modern AI research and applications, finding uses in diverse fields such as image synthesis, art, data augmentation, and even healthcare.

This chapter, "Crafting Images With Generative Adversarial Networks (GANs) and Models," provides a comprehensive guide to understanding and utilizing GANs for image generation. It starts by explaining the fundamental principles and architecture of GANs, including the roles of the generator and discriminator networks. The chapter then progresses into the practical aspects of crafting images, covering data preparation, architectural innovations, training techniques, and evaluation metrics.

IMPORTANCE OF UNDERSTANDING CURRENT CHALLENGES AND FUTURE DIRECTIONS IN THE FIELD OF GANS AND GENERATIVE MODELS

As transformative as GANs have been, the field is still rife with challenges that need to be addressed to unlock their full potential. Understanding these challenges and the future directions of GAN research is crucial for several reasons:

1. ***Enhancing Stability and Performance***: One of the most significant challenges with GANs is their training instability, which often leads to issues like mode collapse and non-convergence. By understanding these problems, researchers can develop more robust training techniques and architectural improvements that enhance the stability and performance of GANs.
2. ***Improving Evaluation Metrics***: Current evaluation metrics for GANs, such as the Inception Score and FID, have their limitations and do not always correlate well with human perception of image quality. Developing better evaluation methods is essential for accurately assessing and comparing the performance of different GAN models, which in turn drives progress in the field.
3. ***Addressing Data Dependency***: GANs require large amounts of high-quality, diverse data to produce realistic outputs. Understanding the impact of data quality and developing methods to train GANs with limited or biased data can make this technology more accessible and effective in various applications.
4. ***Ethical and Societal Implications***: The potential misuse of GANs for creating deepfakes and other malicious purposes poses significant ethical and societal challenges. By understanding these implications, researchers and policymakers can develop guidelines and detection methods to mitigate the risks and ensure responsible use of GANs.

Ethical and Societal Concerns

Solutions for Misuse

The rapid advancements in artificial intelligence, for the most part in Generative Adversarial Networks (GANs), have raised concerns regarding the potential misuse of deepfake technology. To ease these risks, a multi-faceted approach is necessary.

Policy Frameworks and Collaboration

A robust policy framework is essential to regulate the ethical use of AI-generated content. Governments worldwide are working toward implementing laws that specifically address deepfake technology, ensuring responsible AI usage. Collaboration between AI researchers, policymakers, and law enforcement agencies is crucial in drafting regulations that balance innovation and security. Countries such as the United States and the European Union have proposed stringent regulations to combat deepfake-related crimes, including identity fraud and misinformation.

Deepfake Detection Tools

Technological advancements have also contributed to the development of deepfake detection tools. One such tool is **Microsoft's Video Authenticator**, which analyses videos and assigns a confidence score indicating the likelihood of manipulation. Other AI-driven tools, such as Google's Deepfake Detection Challenge models and Facebook's Deepfake Detection Dataset, provide additional resources for identifying synthetic media. The integration of blockchain technology is also being explored to authenticate digital content and prevent unauthorized alterations.

Accountability

Ensuring accountability in AI-generated content is a critical aspect of ethical GAN deployment. The responsibility is shared among various stakeholders, including developers, organizations, and regulatory bodies.

Roles of Developers

Developers play a key role in ensuring that deepfake technology is used ethically. They must adhere to ethical AI principles by designing models that include safeguards against malicious use. Transparent AI development practices, such as embedding watermarks in synthetic content and restricting access to high-risk applications, can help reduce misuse.

Organizational Responsibilities

Organizations utilizing GAN-based technologies must establish ethical guidelines and compliance policies. Tech companies should integrate AI ethics teams to oversee the deployment of generative models, ensuring they align with regulatory standards. Additionally, social media platforms must enhance their content moderation capabilities to detect and remove harmful deepfake content. By combining regulatory policies, technological solutions, and ethical AI practices, it is possible to minimize the risks associated with deepfake misuse while fostering innovation in a responsible manner.

5. ***Future Innovations and Applications***: The future of GANs is bright, with potential integrations with other AI technologies, applications in new domains, and interdisciplinary research offering exciting possibilities. Understanding the current challenges allows researchers to focus on the most impactful areas for innovation, paving the way for new breakthroughs and applications.

SECTION 1: CURRENT CHALLENGES IN GANS

1.1 Training Instability

Training instability is a critical issue in deep learning, often marked by problems such as non-convergence, vanishing gradients, and exploding gradients. Non-convergence occurs when models fail to optimize effectively, often due to poor initialization or incompatible hyperparameters. Vanishing gradients hinder weight updates in deep networks, especially in tasks requiring long-term dependencies, while exploding gradients cause numerical instability during training. Mitigation strategies include the use of advanced architectures and loss functions. Wasserstein GAN (WGAN) addresses instability in generative adversarial networks by using a Wasserstein loss, improving gradient flow and convergence. Spectral Normalization stabilizes training by constraining the Lipschitz constant of the neural network, reducing sensitivity to gradient explosion and ensuring consistent learning dynamics.

Mode collapse, a significant issue in generative models, results in limited diversity of outputs as the generator produces only a subset of possible variations. This undermines the model's capacity to generalize. Improved loss functions, such as those introduced in WGAN-GP, help by stabilizing training and encouraging the generator to explore a broader range of outputs. Evaluation metrics play a crucial role in assessing generative models. While metrics like Inception Score (IS) and Frechet Inception Distance (FID) have been standard, emerging metrics such as Kernel Inception Distance (KID) offer advantages, including computational efficiency and sensitivity to sample quality. A detailed comparative analysis of these metrics highlights their strengths and limitations in capturing fidelity and diversity. Ethical and societal concerns, especially the misuse of generative models in creating deepfakes, pose substantial risks to privacy and misinformation. Addressing these requires robust policy frameworks, improved detection algorithms, and interdisciplinary collaborations to align technological advancements with societal safeguards.

The computational demands of training complex models necessitate efficient use of hardware and software. Optimizing resource usage involves leveraging hardware accelerators like GPUs and TPUs, adopting parallel processing techniques, and employing software optimizations like mixed-precision training to balance performance and cost-effectiveness.

Explanation of Instability Issues in GAN Training

Training instability in GANs arises from the adversarial nature of the generator and discriminator networks. The generator aims to produce data that can fool the discriminator, while the discriminator strives to distinguish between real and generated data. This adversarial process can lead to a delicate balance where small changes in training dynamics can cause instability. Common issues include:

- ***Non-Convergence***: The generator and discriminator may not reach equilibrium, resulting in continuous fluctuations in performance without improving the quality of generated data.
- ***Vanishing Gradients***: If the discriminator becomes too strong, it can easily distinguish between real and fake data, leading to very small gradients for the generator and hindering its learning process.
- ***Exploding Gradients***: Conversely, if the generator becomes too strong, the gradients may become excessively large, causing the model to diverge.

Methods to Mitigate Instability

Researchers have developed several methods to address training instability and mode collapse in GANs. Some of the most effective techniques include:

1. Wasserstein GAN (WGAN)
    I. Concept: WGAN modifies the standard GAN framework by using the Wasserstein distance (also known as Earth Mover's distance) as the loss function. This distance measures the effort required to transform the generated data distribution into the real data distribution.
    II. Benefits: WGAN provides a smoother gradient for the generator, addressing the vanishing gradient problem. It also leads to more stable training by ensuring that the loss function correlates with the quality of the generated data.
    III. Implementation: WGAN requires the discriminator (often called the critic in this context) to satisfy the Lipschitz constraint, which is typically enforced by clipping the weights or using gradient penalty methods.

Illustrative Examples

Visual Aids for Spectral Normalization

Spectral Normalization stabilizes GAN training by limiting the spectral norm of weight matrices, ensuring controlled Lipschitz continuity.

Visual Aids

Incorporating well-structured visual aids enhances comprehension by providing a graphical representation of complex concepts. Figures, diagrams, and illustrations help clarify technical details, making the content more engaging and accessible for readers. To maximize their effectiveness, all figures should include clear and descriptive captions that highlight their relevance to the discussion. Captions not only explain what a figure represents but also reinforce key takeaways from the accompanying text. For instance, when discussing the ***StyleGAN architecture***, a visual diagram can illustrate how different style layers influence the generation of images at various levels of abstraction. A properly labeled diagram would depict how early layers control broad structural details, while later layers refine finer features such as textures and colours. This helps readers understand how StyleGAN manipulates latent space to achieve high-quality image synthesis. Similarly, when explaining ***Progressive Growing***, a diagram can demonstrate how the resolution of images evolves during training. By visualizing the gradual refinement process, from low-resolution blurry outputs to high-resolution, detailed images, readers can better grasp the underlying mechanism. This type of illustration is particularly useful for conveying the advantages of progressive learning in stabilizing GAN training and improving image quality. Beyond architectural diagrams, flowcharts and comparative tables can also enhance understanding by summarizing differences between models, training strategies, and optimization techniques. Thoughtfully designed visuals, accompanied by concise explanations, contribute to a well-structured and informative chapter.

1.2 Evaluation Metrics

Evaluation Metrics in GANs

Evaluating the performance of Generative Adversarial Networks (GANs) is a complex and challenging task. Unlike traditional machine learning models that can be evaluated using straightforward metrics like accuracy or precision, GANs generate data, requiring different approaches to assess their quality and realism. This segment discusses the difficulties in evaluating GANs, outlines common metrics used in the field, and highpoints the limitations of these metrics.

Difficulty in Evaluating GANs

The evaluation of GANs is inherently difficult due to several factors:

1. ***Subjectivity of Quality***: The quality of generated images is often subjective, and what looks realistic to one observer might not appear as convincing to another.
2. ***Lack of Ground Truth***: In many cases, there is no single "correct" output for GANs to compare against, making it hard to quantify the accuracy of the generated data.
3. ***Diversity and Mode Collapse***: Evaluating whether a GAN captures the full diversity of the training data distribution is challenging, especially in detecting mode collapse where the generator produces limited variations of data.

Limitations of Current Metrics

Despite their widespread use, both the Inception Score and Fréchet Inception Distance have notable limitations:

1. Inception Score (IS)
    - ***Sensitivity to Classifier***: IS relies on a specific pre-trained classifier (Inception v3) and its performance can vary depending on how well the classifier's training data aligns with the GAN's output domain.
    - ***Lack of Diversity Measurement***: IS primarily focuses on the confidence of individual image classifications and does not explicitly measure diversity within the generated images, potentially overlooking mode collapse.
2. Fréchet Inception Distance (FID)
    - ***Dependence on Pre-Trained Model***: Similar to IS, FID relies on features extracted by a pre-trained Inception v3 model, which may not be optimal for all types of generated images, especially those that deviate significantly from the training data of the Inception model.
    - ***Computational Complexity***: FID requires calculating the mean and covariance of high-dimensional feature vectors, which can be computationally expensive and sensitive to the number of samples used.
    - ***Statistical Sensitivity***: FID can be sensitive to small sample sizes and noise in the data, potentially leading to misleading evaluations.

1.3 Data Dependency and Quality in GANs

Generative Adversarial Networks (GANs) are heavily dependent on the quality and diversity of the data used to train them. The performance and generalization ability of GANs are directly influenced by the nature of the training dataset. High-quality, well-balanced datasets enable GANs to generate realistic and diverse outputs, while poor-quality or biased data can severely limit their effectiveness and applicability. The impact of data quality on GAN performance cannot be overstated. When trained on high-quality datasets, GANs learn to generate images that closely mimic the intricacies and details of real-world data. These datasets typically encompass a wide variety of examples, providing the GAN with a comprehensive understanding of the target distribution. As a result, the generator produces outputs that are not only realistic but also varied, capturing the richness of the original dataset. On the other hand, if the dataset is of poor quality—containing noisy, low-resolution, or inconsistent data—the GAN struggles to learn meaningful features. This often leads to generated images that are blurry, lack detail, or exhibit artifacts. Furthermore, a dataset with limited diversity can cause mode collapse, where the generator produces a narrow range of outputs, failing to capture the full spectrum of the target distribution. This undermines the GAN's ability to generalize, making it less effective in real-world applications where diversity and quality are crucial.

Challenges arise not only from the quality of the data but also from its sufficiency and balance. Insufficient data can lead to overfitting, where the GAN memorizes the training examples rather than learning to generalize. This is particularly problematic in scenarios where obtaining large datasets is difficult, such as in medical imaging or niche artistic domains. Researchers have attempted to mitigate this by using data augmentation techniques, which artificially expand the dataset by applying transformations to existing images. However, these techniques can only go so far in providing the diversity needed for robust GAN performance. Biased data presents another significant challenge. If the training dataset is biased—whether due to demographic imbalances, cultural stereotypes, or other forms of skewed representation—the GAN will learn and perpetuate these biases in its generated outputs.

1.4 Computational Resources

Training Generative Adversarial Networks (GANs) requires substantial computational resources, posing significant challenges in terms of cost and scalability. The process involves iterative optimization of two neural networks—the generator and the discriminator—through a complex adversarial training process. This typically demands extensive computational power, often necessitating the use of high-performance GPUs or specialized hardware like TPUs. These requirements can lead to high financial costs, making GAN research and application development expensive, particularly for smaller organizations or individual researchers.

The high computational costs are driven by several factors. Training GANs involves numerous epochs to reach convergence, especially when dealing with large and complex datasets. Additionally, advanced GAN architectures, such as StyleGAN or BigGAN, have millions of parameters, increasing the demand for memory and processing power. The iterative nature of adversarial training further exacerbates these requirements, as each step involves calculating gradients and updating the networks, which is computationally intensive. Access to these computational resources is a significant barrier for many. While large tech companies and well-funded research institutions can afford the necessary infrastructure, smaller entities may struggle to compete. This disparity limits the democratization of GAN technology, hindering widespread experimentation and innovation. Scalability also becomes an issue as the complexity and size of models increase. Efficiently training large-scale GANs on distributed systems requires sophisticated techniques to manage communication overhead and ensure synchronized updates, adding another layer of complexity. The high computational costs and resource accessibility issues present substantial obstacles in the development and deployment of GANs. Addressing these challenges is crucial for making GAN technology more accessible and scalable, enabling broader participation and fostering innovation in the field.

1.5 Ethical and Societal Concerns

The rise of Generative Adversarial Networks (GANs) has brought significant ethical and societal concerns, particularly with deepfakes and synthetic media. Deepfakes, which are hyper-realistic AI-generated videos and images, pose risks such as misinformation, privacy invasion, and reputational damage. These sophisticated forgeries can spread false information, manipulate public opinion, and undermine trust in digital content. Additionally, bias and fairness in GAN-generated content are critical issues. GANs trained on biased datasets can perpetuate and even amplify existing societal biases, leading to unfair and discriminatory outputs. This is particularly concerning in applications like facial recognition and hiring processes, where biased outputs can have real-world negative impacts.

The potential misuse of GANs extends beyond deepfakes. They can be used for malicious purposes, such as creating deceptive media for fraud, political manipulation, or spreading harmful stereotypes. Addressing these ethical challenges is essential to ensure the responsible and fair use of GAN technology.

SECTION 2: ADVANCES AND SOLUTIONS

2.1 Improved Architectures

Architectural innovations in Generative Adversarial Networks (GANs) have significantly advanced the field, with models like StyleGAN and BigGAN leading the charge. StyleGAN introduces a novel approach to control the style of generated images at different levels of detail. By incorporating adaptive instance normalization and a progressive growing strategy, StyleGAN enhances the quality and resolution of images, addressing issues of training instability and mode collapse. This architecture allows for precise manipulation of image attributes, producing highly realistic and diverse outputs. BigGAN, on the other hand, focuses on scaling up GANs to generate high-fidelity images. It leverages larger batch sizes, deeper architectures, and orthogonal regularization to stabilize training and improve the diversity of generated samples. BigGAN's approach mitigates the problem of mode collapse by ensuring a more robust representation of the data distribution.

These architectural improvements directly tackle the core challenges of GANs, such as training instability, limited diversity, and poor image quality. By refining the network structures and incorporating innovative training techniques, StyleGAN and BigGAN push the boundaries of what GANs can achieve, enabling the generation of more realistic and varied images.

2.2 Enhanced Training Techniques

Enhanced training techniques have played a crucial role in advancing the stability and performance of Generative Adversarial Networks (GANs). Methods such as progressive growing and curriculum learning have significantly contributed to these improvements. Progressive growing involves gradually increasing the size and complexity of the GAN architecture during training. Initially, both the generator and discriminator are trained on low-resolution images. As training progresses, new layers are added to increase the image resolution incrementally. This approach stabilizes training by allowing the model to learn basic features at lower resolutions before focusing on finer details, reducing the risk of instability and mode collapse. Progressive growing has been particularly effective in improving image quality and enabling the generation of high-resolution images. Curriculum learning applies a strategy where the GAN is trained on simpler tasks before gradually advancing to more complex ones. By starting with easier objectives and progressively introducing more challenging scenarios, the GAN can learn in a more stable and controlled manner. This technique helps to manage the adversarial dynamics between the generator and discriminator, reducing the likelihood of training failures and enhancing the overall robustness of the model. The impact of these advanced training methods is substantial, resulting in more stable GAN training processes and improved performance. They facilitate the generation of higher quality and more diverse images, making GANs more reliable and effective for various applications.

2.3 Better Evaluation Methods

Developing more reliable and comprehensive evaluation metrics is crucial for accurately assessing Generative Adversarial Networks (GANs). Current metrics like Inception Score (IS) and Fréchet Inception Distance (FID) have limitations, prompting the need for better methods. Researchers are exploring new metrics that capture both the quality and diversity of generated images more effectively. These metrics aim to provide a holistic view of GAN performance, addressing aspects like perceptual quality and realism. Incorporating human judgment remains essential, as humans can evaluate subtleties and nuances that automated metrics might miss. Hybrid approaches, combining quantitative metrics with qualitative human assessments, offer a more balanced and thorough evaluation. This combination ensures that the GAN-generated content not only meets statistical benchmarks but also aligns with human perceptions of quality and realism, leading to more accurate and meaningful assessments of GAN performance.

Generative Adversarial Networks (GANs) have significantly advanced the field of artificial intelligence by enabling the creation of high-quality synthetic data. However, evaluating the performance of GANs has proven to be challenging due to the subjective nature of generated outputs and the lack of standardized metrics. Traditional evaluation methods often fall short in capturing the nuanced aspects of GAN performance, necessitating the development of better evaluation techniques. This essay explores various advanced methods for evaluating GANs, focusing on improvements in both quantitative and qualitative assessments.

2.3.1 Overview of Traditional Evaluation Methods

Before delving into advanced methods, it's essential to understand traditional evaluation approaches:

Inception Score (IS): Measures the quality of generated images based on the classification confidence of an Inception network. A high IS indicates that images are easily classified into distinct categories. However, IS has limitations, including a lack of sensitivity to mode collapse and a reliance on the classifier's performance.

Frechet Inception Distance (FID): Evaluates the quality of generated images by comparing the distribution of features extracted from a pre-trained Inception network. FID computes the distance between the feature distributions of real and generated images, offering a more comprehensive measure than IS. Despite its advantages, FID can be sensitive to changes in the network architecture and may not fully capture diversity.

Precision and Recall: Precision measures the proportion of generated samples that are real-like, while recall assesses the proportion of real data that the generator is able to capture. These metrics offer insights into both the quality and diversity of generated samples but can be complex to compute and interpret.

2.3.2 Advanced Quantitative Metrics

To address the limitations of traditional metrics, several advanced quantitative evaluation methods have been proposed:

a. Perceptual Path Length (PPL)

Description: PPL measures the smoothness of the latent space of GANs by evaluating the perceptual distance between images generated from nearby points in the latent space. A low PPL indicates that small changes in the latent space result in meaningful changes in generated images, suggesting a well-structured latent space.

Advantages: Provides insights into the quality of the latent space representation and the smoothness of the image generation process.

Challenges: Requires careful calibration and can be computationally expensive to evaluate.

b. Kernel Inception Distance (KID)

Description KID is a variation of FID that uses polynomial kernels to compare feature distributions. It is designed to be less sensitive to the number of samples and provides a more reliable measure of distributional similarity.

Advantages: Addresses some of the sensitivity issues of FID and is robust to the choice of feature extractor.

Challenges: Kernel methods can be computationally intensive and may require optimization of kernel parameters.

c. Mode Score

Description: Mode Score evaluates both the quality and diversity of generated samples by measuring the density of generated samples within each mode of the data distribution. It addresses mode collapse by penalizing generators that produce fewer modes.

Advantages: Provides a balanced measure of quality and diversity, mitigating the effects of mode collapse.

Challenges: Requires accurate estimation of mode density and can be sensitive to the choice of clustering algorithm.

d. Multi-scale Structural Similarity Index (MS-SSIM)

Description: MS-SSIM extends the traditional SSIM metric by evaluating image quality at multiple scales, offering a more comprehensive assessment of structural similarity between real and generated images.

Advantages: Captures finer details and structural information, providing a more nuanced measure of image quality.

Challenges: Can be computationally expensive and may require optimization of scale parameters.

2.3.3 Qualitative Evaluation Techniques

Quantitative metrics alone may not fully capture the quality of generated content, making qualitative evaluation techniques essential:

a. Human Evaluation

Description: Involves subjective assessments by human judges who evaluate the quality and realism of generated samples. This method can include tasks such as rating images or identifying which images are real or fake.

Advantages: Provides direct feedback on perceptual quality and realism, capturing aspects that metrics may miss.

Challenges: Subjective nature can introduce variability, and scaling human evaluations can be resource-intensive.

b. User Studies

Description: Involves conducting studies where users interact with or use generated content in real-world scenarios. This can provide insights into the practical utility and effectiveness of the generated content.

Advantages: Offers practical insights into the usability and effectiveness of GAN-generated content in real-world applications.

Challenges: Requires careful study design and may be difficult to generalize across different applications.

c. Visual Turing Test

Description: A variation of the Turing Test where human judges are asked to distinguish between real and generated images. The performance of judges can provide an indication of the realism of generated content.

Advantages: Directly measures the ability of generated content to deceive human observers, providing a clear assessment of realism.
Challenges: The effectiveness of the test can depend on the quality of the generated content and the expertise of the judges.

2.3.4 Evaluation Frameworks and Metrics

To systematically assess GAN performance, several evaluation frameworks and metrics have been proposed:

a. Generative Model Evaluation Framework (GMEF)
Description: GMEF provides a comprehensive framework for evaluating generative models, encompassing multiple metrics and evaluation criteria, including quality, diversity, and usability.
Advantages: Offers a holistic approach to evaluation, integrating various metrics and qualitative assessments.
Challenges: Requires careful integration of different evaluation aspects and may be complex to implement.

b. Evaluation Scoreboards
Description: Online platforms or tools that aggregate and compare the performance of different GAN models across various metrics and tasks.
Advantages: Facilitates benchmarking and comparison, providing a standardized way to evaluate and compare GAN performance.
Challenges Requires regular updates and maintenance to ensure relevance and accuracy.

c. Visualization Techniques
Description: Involves using advanced visualization techniques, such as t-SNE or UMAP, to explore and analyze the latent space and generated samples. These techniques can help identify patterns and anomalies in generated content.
Advantages: Provides visual insights into the structure of generated content and latent space, aiding in understanding and debugging GAN performance.
Challenges: Interpretation of visualizations can be subjective and may require expertise in visualization techniques.

2.3.5 Addressing Challenges and Future Directions

Despite advances in evaluation methods, several challenges remain:

Standardization: There is a need for standardized evaluation practices to ensure consistency and comparability across different GAN models and tasks.
Complexity: Evaluating GANs often requires a combination of metrics and qualitative assessments, making the evaluation process complex and resource-intensive.
Scalability: As GAN models become more advanced, evaluation methods must scale to handle larger datasets and more complex outputs.

2.4 Resource Optimization

Resource optimization in GAN training focuses on reducing computational costs while maintaining performance. Techniques like model compression and efficient GANs play a significant role in this area. Model compression involves methods such as pruning, quantization, and knowledge distillation to reduce the size and complexity of GAN models. These approaches lower memory usage and speed up computation without significantly compromising the quality of generated images. Efficient GANs, such as MobileGAN and LiteGAN, are designed with streamlined architectures that require fewer resources. These models employ techniques like layer reduction, optimized activation functions, and enhanced training protocols to achieve high-quality outputs more efficiently. Advances in hardware and parallel processing further enhance resource optimization. Modern GPUs and TPUs offer increased processing power and memory capabilities, enabling faster training times and handling larger models. Additionally, parallel processing techniques, including distributed computing and multi-GPU setups, allow for more efficient utilization of resources, significantly reducing the overall computational burden. These advancements collectively enable more accessible and scalable GAN training and deployment.

2.5 Addressing Ethical Issues

Addressing ethical issues in GANs involves implementing strategies to mitigate risks of misuse and establishing frameworks for responsible development. To counter the spread and impact of deepfakes, researchers have developed advanced detection techniques. These techniques utilize machine learning algorithms to identify subtle artifacts and inconsistencies in synthetic media that are not easily perceptible to the human eye. By continuously updating these detection methods to keep pace with evolving GAN capabilities, it becomes possible to effectively distinguish between real and manipulated content. Establishing robust frameworks for the ethical use and development of GANs is equally important. These frameworks include guidelines and best practices for researchers and developers to ensure fairness, transparency, and accountability. Principles such as data privacy, consent, and bias mitigation are integral to these frameworks, promoting the development of GANs in ways that minimize harm and maximize societal benefits. Ethical oversight and collaboration with interdisciplinary experts further reinforce these frameworks, ensuring that the advancements in GAN technology are aligned with broader societal values and norms.

Generative Adversarial Networks (GANs) have significantly advanced the field of artificial intelligence (AI) by enabling the creation of highly realistic synthetic data. Despite their revolutionary impact, GANs pose several ethical challenges that must be addressed to ensure their responsible and beneficial use. This composition explores these ethical issues in depth, including concerns related to misinformation, privacy, consent, intellectual property, bias, security, and societal impact, and discusses potential strategies for mitigating these concerns.

2.6 Language and Grammar

Ensuring clarity and readability is essential when refining technical writing. Simplifying sentences and using an active voice improves the flow of information, making it easier for readers to grasp complex concepts. Passive constructions often make sentences longer and more difficult to follow, while active voice helps to present ideas more directly. For example, a sentence likes "The training of GANs often suffers from instability, including vanishing gradients. "Can be revised to "GAN training often suffers from instability, including vanishing gradients." This version eliminates unnecessary words, making the statement more concise while maintaining its original meaning.

In addition to sentence structure, maintaining consistency in terminology enhances the coherence of a research paper. Variations in phrasing, such as using "GAN-generated content" in one section and "content generated by GANs" in another, may cause slight inconsistencies. Standardizing terms throughout the document ensures a more polished and professional presentation. Proofreading also plays a critical role in refining language quality. Careful review of grammar, punctuation, and technical accuracy helps prevent misinterpretations and strengthens the credibility of the research. Consistently applying proper terminology and structured sentence formation ensures that the manuscript adheres to academic writing standards while remaining accessible to the intended audience.

SECTION 3: FUTURE DIRECTIONS IN GAN RESEARCH

Future Directions

The future of Generative Adversarial Networks (GANs) lies in their integration with other AI technologies, enhancing their adaptability and expanding their applications across various domains. As research in artificial intelligence progresses, GANs are expected to work alongside other machine learning models to improve efficiency, realism, and accuracy in generated data. One promising direction is the combination of GANs with reinforcement learning, which allows for more adaptive and dynamic data generation. This integration is particularly useful in areas such as autonomous driving simulations, where realistic synthetic environments can be created to train self-driving algorithms. By continuously refining the training process through feedback loops, reinforcement learning helps GANs generate high-quality, scenario-specific datasets, ultimately improving decision-making capabilities in autonomous systems. In addition to reinforcement learning, transformers—known for their success in natural language processing—are also being integrated with GANs to enhance the generation of coherent and contextually accurate content. Transformers improve text and image synthesis by ensuring logical flow and semantic consistency. This is especially valuable in applications such as content creation, automated storytelling, and AI-driven design, where the generated material needs to align with human-like reasoning and context.

Observing forward, GANs are expected to drive significant advancements in fields like personalized medicine and education. In healthcare, GANs can be used to generate patient-specific synthetic medical images, allowing doctors to simulate disease progression and tailor treatment plans accordingly. This personalized approach to medicine has the potential to enhance diagnostics and drug development, making treatments more effective and targeted. Similarly, in education, GANs can contribute to the development of intelligent tutoring systems that adapt to individual learning styles. By generating personalized educational content, GANs can help create interactive and engaging learning experiences, making complex subjects more accessible to students. These AI-powered educational tools have the potential to transform traditional learning methodologies by offering customized support based on a learner's unique needs.

3.1 Integration with Other AI Technologies

Integrating GANs with other AI technologies, such as reinforcement learning and transformers, can unlock new capabilities and applications, enhancing the overall performance and versatility of these models.

Reinforcement learning (RL), when combined with GANs, can facilitate more sophisticated and adaptive data generation. For example, RL can guide the generator in a GAN to produce outputs that meet specific criteria or adapt to dynamic environments. This integration is particularly useful in scenarios like autonomous driving, where realistic simulations are crucial for training and testing algorithms.

Transformers, known for their prowess in handling sequential data and capturing long-range dependencies, complement GANs by enhancing their ability to generate coherent and contextually accurate outputs. In natural language processing (NLP), combining transformers with GANs can improve text generation, creating more fluent and contextually relevant sentences. Similarly, in image synthesis, transformers can help in generating images with better structural coherence and detail.

3.2 Applications in New Domains

GANs are making significant inroads into new domains such as healthcare, art, and education, demonstrating their versatility and potential for innovation.

Healthcare: In medical imaging, GANs enhance diagnostic accuracy by generating high-quality images from low-resolution scans, aiding in early detection and diagnosis. For instance, GANs can create synthetic MRI scans to augment training datasets, improving the performance of diagnostic algorithms.

Art: GANs are revolutionizing the art world by enabling the creation of new, unique pieces. Artists use GANs to explore novel styles and generate artwork that blends different techniques. A notable case is the "Edmond de Belamy" portrait, generated by a GAN, which sold at auction for a significant sum, highlighting the acceptance of AI-generated art in mainstream culture.

Education: In education, GANs enhance learning experiences by creating realistic simulations and interactive content. They can generate diverse datasets for training purposes, especially in fields requiring practical skills, such as virtual labs for scientific experiments or realistic scenarios for language learning.

3.3 Advancements in Conditional GANs

Advancements in Conditional GANs (cGANs) are revolutionizing personalized content generation by enabling more precise and user-specific outputs. cGANs extend the traditional GAN framework by incorporating additional information, or conditions, which guide the generation process. This approach allows for outputs tailored to specific requirements, making cGANs particularly valuable in fields such as healthcare, marketing, and entertainment.

The future of cGANs in personalized content generation looks bright. In healthcare, cGANs can create patient-specific medical images, enhancing diagnostic accuracy and treatment personalization. For instance, a cGAN could generate customized MRI scans based on individual patient data, improving early detection and tailored intervention strategies. In marketing, cGANs can generate personalized advertisements by analyzing user preferences and behaviors, leading to more engaging and effective campaigns. In entertainment, cGANs can create custom media content, such as personalized avatars, music, or video recommendations, enhancing user experience and satisfaction. Innovative conditioning mechanisms are key to these advancements. Text-to-image synthesis, for example, allows cGANs to generate images based on textual descriptions, facilitating applications in personalized graphic design and interactive storytelling. Multimodal conditioning, which combines text, audio, and images, creates richer, more immersive content experiences. These mechanisms enhance the flexibility and applicability of cGANs, making them powerful tools for generating highly specific and personalized outputs. Global, the advancements in cGANs are driving significant progress in personalized content generation, promising exciting developments across various sectors.

3.4 Interdisciplinary Research

Interdisciplinary Research in GANs

Interdisciplinary research, involving collaboration between computer science, psychology, and other fields, has a profound impact on the understanding and improvement of Generative Adversarial Networks (GANs). This collaborative approach leverages diverse expertise to address the multifaceted challenges of GAN development and application.

Computer Science and Machine Learning: At the core, computer science drives the technical advancements in GAN architectures, algorithms, and optimization techniques. Researchers develop novel methods to enhance training stability, improve image quality, and expand the applicability of GANs across various domains.

Psychology and Cognitive Science: Psychology contributes to the understanding of how humans perceive and evaluate generated content. Insights from cognitive science help in designing GANs that produce more realistic and appealing outputs by aligning with human perceptual processes. For instance, studies on visual cognition can inform the development of GANs that better mimic human-like features in generated images, improving their realism.

Ethics and Social Sciences: Interdisciplinary collaboration extends to ethics and social sciences, addressing the societal implications of GANs. Ethical frameworks guide the responsible development and deployment of GANs, ensuring fairness, transparency, and accountability. Social scientists explore the impact of GAN-generated content on society, such as the effects of deepfakes on public trust and the role of AI in perpetuating biases.

Art and Design: Collaboration with artists and designers enhances the creative potential of GANs. By integrating artistic perspectives, GANs can be used to generate novel artworks, pushing the boundaries of digital creativity. This intersection also explores how GANs can assist in creative processes, offering tools for artists to experiment and innovate.

Healthcare and Medical Sciences: In healthcare, interdisciplinary research combines computer science with medical expertise to develop GAN applications for diagnostics, treatment planning, and medical imaging. GANs can generate synthetic medical data for training and testing, aiding in the development of robust AI models for healthcare.

Impact on Understanding and Improving GANs

Interdisciplinary research fosters a holistic approach to improving GANs by addressing both technical and human-centric aspects. The integration of diverse perspectives leads to the development of GANs that are not only technically advanced but also ethically sound and socially beneficial.

- ***Enhanced Realism and Utility***: By incorporating knowledge from psychology and cognitive science, GANs can produce outputs that align more closely with human perception, enhancing their realism and utility in applications like virtual reality, gaming, and digital media.
- ***Ethical and Responsible AI***: Ethical guidelines and social science research ensure that GAN development considers the broader societal impact, pro-

moting fairness and mitigating biases. This results in more trustworthy and socially responsible AI systems.

- ***Innovative Applications***: Collaborations with various fields spur innovative applications of GANs, from personalized healthcare solutions to creative tools for artists, expanding the horizons of what GANs can achieve.
- ***Improved Training and Evaluation***: Insights from interdisciplinary research inform better training techniques and evaluation metrics, addressing challenges like training instability and subjective quality assessment, leading to more robust and reliable GANs.

3.5 Open Challenges

Generative Adversarial Networks (GANs) have achieved impressive milestones, yet several unresolved issues and open questions persist in the field. One of the primary challenges is training stability. GANs often face difficulties during training due to problems such as mode collapse, vanishing gradients, and non-convergence. Although techniques like Wasserstein GAN and spectral normalization have provided improvements, a universally reliable method for training GANs across diverse applications and datasets is still lacking. Mode collapse remains a significant hurdle, where the generator produces a limited variety of outputs, failing to capture the full diversity of the data distribution. Ensuring that GANs generate diverse samples representative of the entire data distribution is critical for their effectiveness and generalization. Additionally, current evaluation metrics like Inception Score (IS) and Fréchet Inception Distance (FID) have their limitations and do not always align with human judgments of quality and diversity. Developing more comprehensive and reliable evaluation metrics is essential for accurately assessing GAN performance. Bias and fairness are also pressing concerns. GANs can perpetuate and amplify biases present in the training data, leading to unfair and discriminatory outputs. Addressing this issue requires the development of methods to detect and mitigate biases, ensuring fairness and inclusivity in GAN-generated content. Moreover, the resource intensity of training GANs poses a challenge, as it demands substantial computational power and hardware resources. Optimizing GAN architectures and training processes to be more efficient without sacrificing performance is an ongoing necessity.

Ethical concerns are paramount, particularly regarding the potential misuse of GANs for creating deepfakes and other harmful applications. Developing robust detection methods and establishing ethical guidelines for the use of GANs are critical steps in preventing abuse and ensuring the responsible deployment of this technology. Despite these challenges, the field of GANs offers numerous opportunities for future research and exploration. Innovating in training techniques is crucial. Researchers can explore novel algorithms and optimization methods to enhance training stability and mitigate mode collapse, such as adaptive learning rates, advanced regularization techniques, and hybrid models.

Case Studies

StyleGAN: Application in Creating Photorealistic Portraits and Artistic Experiments

StyleGAN is a generative model known for producing high-quality, photorealistic human portraits. It allows fine control over various features such as facial expressions, hair texture, and background through its unique style-based architecture. Beyond portraits, StyleGAN has been widely used for artistic experiments, creating surreal digital art by blending different styles or generating entirely new faces. This model is revolutionizing industries like video games, fashion, and digital art by enabling the creation of highly customizable and realistic virtual characters.

Applications:

- **Virtual Characters in Video Games and Movies**: Creating digital human models with precise control over their features.
- **Fashion and Apparel**: Designers use StyleGAN to generate images of models wearing new clothing styles.
- **Art Installations**: Artists have used StyleGAN to create unique, ever-evolving portraits as part of modern digital art.

BigGAN: Large-Scale Natural Image Synthesis for Training Datasets

BigGAN is designed for large-scale, high-resolution natural image synthesis, producing diverse and realistic images of objects, animals, and scenes. It is especially useful for generating large datasets to train deep learning models for tasks like object detection, autonomous driving, and medical imaging. By creating high-quality synthetic data, BigGAN helps overcome challenges in acquiring labeled data and augments datasets for improved model performance, enhancing industries reliant on computer vision and AI-based systems.

Applications:

- **Training Datasets for Computer Vision Models**: BigGAN has been particularly valuable in generating large-scale synthetic datasets for training computer vision models. For example, datasets of natural scenes can be generated to train models for object detection, segmentation, and classification tasks.

- **Autonomous Vehicles**: Synthetic data generated by BigGAN can be used to train machine learning models for self-driving cars, providing simulated images of roads, traffic signs, and pedestrians.
- **Medical Imaging**: In fields like radiology, where labeled datasets are often limited, BigGAN can generate synthetic medical images (e.g., X-rays, MRIs) that assist in training diagnostic models.
- **Augmenting Existing Datasets**: BigGAN-generated images can supplement real-world datasets in cases where data augmentation is necessary, enhancing model robustness and generalization.

CONCLUSION

Generative Adversarial Networks (GANs) represent one of the most transformative advancements in artificial intelligence, with their ability to generate realistic images, texts, and other forms of data. However, despite their impressive capabilities, GANs face several key challenges that need to be addressed to unlock their full potential. One of the primary challenges is training stability. GANs often struggle with issues such as mode collapse, vanishing gradients, and non-convergence, making the training process difficult and unpredictable. Techniques like Wasserstein GAN and spectral normalization have offered improvements, but a universally reliable method for stabilizing GAN training across various applications and datasets remains elusive. Mode collapse is another significant hurdle. This issue arises when the generator produces a limited variety of outputs, failing to capture the full diversity of the data distribution. Overcoming mode collapse is crucial for ensuring that GANs can generate a wide range of samples that accurately represent the underlying data. Evaluation metrics are also a critical area of concern. Current metrics, such as Inception Score (IS) and Fréchet Inception Distance (FID), have limitations and do not always correlate well with human judgments of quality and diversity. Developing more comprehensive and reliable evaluation metrics is essential for accurately assessing the performance of GANs and guiding their development. Bias and fairness present additional challenges. GANs trained on biased datasets can perpetuate and even amplify existing biases, leading to unfair and discriminatory outputs. Addressing these issues requires the development of methods to detect and mitigate biases, ensuring that GAN-generated content is fair and inclusive. The high resource intensity of training GANs is another significant challenge. Training GANs typically requires substantial computational power and hardware resources, which can be a barrier to their widespread adoption and application. Optimizing GAN architectures and training processes to be more efficient is an ongoing necessity. Ethical concerns are paramount, particularly regarding the potential misuse of GANs for creating deepfakes and other harmful applications. Developing robust detection methods and establishing ethical guidelines for the responsible use of GANs are critical steps in preventing abuse and ensuring that this technology is deployed in ways that benefit society. Despite these challenges, the advancements in GAN technology are impressive and promising. Innovations such as StyleGAN and BigGAN have significantly improved the quality and realism of generated images. Enhanced training techniques, including progressive growing and curriculum learning, have contributed to greater stability and performance in GAN training. The integration of GANs with other AI technologies, such as reinforcement learning and transformers, has expanded their capabilities and applications.

1. Looking to the future, the vision for GANs and generative models is one of continued evolution and expansion. Addressing the current challenges through innovative research, ethical considerations, and cross-disciplinary collaboration will be key to driving the next wave of advancements. Researchers are exhilarated to explore novel algorithms and optimization methods to enhance training stability and mitigate mode collapse.
2. Developing advanced evaluation metrics that better capture the quality and diversity of GAN outputs will provide more robust assessment frameworks. Focusing on fairness and ethics will be crucial as GAN technology becomes more integrated into various applications. Methods to identify and reduce biases in training data and generated content, along with comprehensive ethical guidelines, will ensure the responsible use of GANs.
3. Enhancing efficiency through research on model compression, efficient architectures, and advanced hardware utilization will make GAN training more accessible and scalable. Interdisciplinary collaboration involving computer science, psychology, ethics, healthcare, and other fields will provide holistic solutions and innovative applications for GANs. Exploring new applications in novel domains, such as environmental modelling, education, and personalized content generation, will demonstrate the potential and versatility of GANs, paving the way for broader adoption and impact.
4. The evolution and potential of GAN technology are vast and exciting. While significant challenges remain, the advancements and ongoing research efforts are paving the way for GANs to become more robust, fair, and widely usable tools in various fields. By addressing these challenges and continuing to innovate, the future of GANs and generative models holds great promise for transforming industries and improving our understanding and manipulation of data in unprecedented ways.

5. Generative Adversarial Networks (GANs) have emerged as a transformative technology with applications spanning multiple industries, from healthcare and finance to creative arts and entertainment. Though challenges such as training instability, ethical concerns, and computational costs must be addressed to unlock their full potential. By overcoming these limitations, GANs can drive significant advancements, enabling more accurate medical diagnostics, realistic content generation, and intelligent automation. One of the most pressing concerns in GAN research is **bias in datasets**, which can lead to unfair or misleading AI-generated outputs. Addressing this issue requires curating diverse and representative training data, implementing fairness-aware algorithms, and continually evaluating model outputs for unintended biases.

REFERENCES

Arjovsky, M., Chintala, S., & Bottou, L. (2017). Wasserstein GAN. arXiv preprint arXiv:1701.07875. https://doi.org//arXiv.1701.07875DOI: 10.48550

Brock, A., Donahue, J., & Simonyan, K. (2019). Large scale GAN training for high fidelity natural image synthesis. arXiv preprint arXiv:1809.11096. https://doi.org//arXiv.1809.11096DOI: 10.48550

Brock, A., Donahue, J., & Simonyan, K. (2019). Large scale GAN training for high fidelity natural image synthesis. *International Conference on Learning Representations (ICLR)*. https://doi.org/DOI: 10.48550/arXiv.1809.11096

Brock, A., Donahue, J., & Simonyan, K. (2019). Large Scale GAN Training for High Fidelity Natural Image Synthesis. *International Conference on Learning Representations (ICLR)*. https://doi.org/DOI: 10.48550/arXiv.1809.11096

Chen, X., Duan, Y., Houthooft, R., Schulman, J., Sutskever, I., & Abbeel, P. (2016). InfoGAN: Interpretable representation learning by information maximizing generative adversarial nets. Advances in Neural Information Processing Systems (NeurIPS), 29, 2172-2180. https://doi.org/DOI: 10.48550/arXiv.1606.03657

Creswell, A., Bharath, A. A., & Bharath, A. (2018). Generative adversarial networks: An overview. *IEEE Signal Processing Magazine*, 35(1), 53–65. DOI: 10.1109/MSP.2017.2765202

Denton, E. L., Chintala, S., Szlam, A., & Fergus, R. (2015). Deep generative image models using a Laplacian pyramid of adversarial networks. Advances in Neural Information Processing Systems (NeurIPS), 1486-1494. https://doi.org/DOI: 10.48550/arXiv.1506.05751

Ghosh, A., Kulharia, V., Mukherjee, S., Dokania, P. K., Torr, P. H. S., & Sminchisescu, C. (2020). From variational to deterministic autoencoders: Understanding reconstruction via Wasserstein estimation. *Proceedings of the IEEE/CVF Conference on Computer Vision and Pattern Recognition (CVPR)*, 11541-11550. https://doi.org/DOI: 10.1109/CVPR42600.2020.01156

Goodfellow, I., Pouget-Abadie, J., Mirza, M., Xu, B., Warde-Farley, D., Ozair, S., & Bengio, Y. (2014). Generative adversarial nets. *Advances in Neural Information Processing Systems*, 27, 2672–2680. DOI: 10.48550/arXiv.1406.2661

Goodfellow, I., Pouget-Abadie, J., Mirza, M., Xu, B., Warde-Farley, D., Ozair, S., Courville, A., & Bengio, Y. (2014). Generative Adversarial Networks. Advances in Neural Information Processing Systems (NeurIPS), 27. https://doi.org/DOI: 10.48550/arXiv.1406.2661

Huang, X., & Belongie, S. (2017). Arbitrary style transfer in real-time with adaptive instance normalization. *Proceedings of the IEEE International Conference on Computer Vision (ICCV)*, 1501-1510. https://doi.org/DOI: 10.1109/ICCV.2017.167

Isola, P., Zhu, J. Y., Zhou, T., & Efros, A. A. (2017). Image-to-image translation with conditional adversarial networks. *IEEE Conference on Computer Vision and Pattern Recognition*, 1125-1134. https://doi.org/DOI: 10.1109/CVPR.2017.632

Jolicoeur-Martineau, A. (2019). The relativistic discriminator: A key element missing from standard GAN. *International Conference on Learning Representations (ICLR)*. https://doi.org/DOI: 10.48550/arXiv.1807.00734

Karras, T., Aila, T., Laine, S., & Lehtinen, J. (2018). Progressive growing of GANs for improved quality, stability, and variation. arXiv preprint arXiv:1710.10196. https://doi.org//arXiv.1710.10196DOI: 10.48550

Karras, T., Aila, T., Laine, S., & Lehtinen, J. (2018). Progressive Growing of GANs for Improved Quality, Stability, and Variation. *International Conference on Learning Representations (ICLR)*. https://doi.org/DOI: 10.48550/arXiv.1710.10196

Karras, T., Aila, T., Laine, S., & Lehtinen, J. (2018). Progressive growing of GANs for improved quality, stability, and variation. *International Conference on Learning Representations (ICLR)*. https://doi.org/DOI: 10.48550/arXiv.1710.10196

Karras, T., Laine, S., & Aila, T. (2019). A style-based generator architecture for generative adversarial networks. *IEEE/CVF Conference on Computer Vision and Pattern Recognition*, 4401-4410. https://doi.org/DOI: 10.1109/CVPR.2019.00453

Karras, T., Laine, S., & Aila, T. (2019). A Style-Based Generator Architecture for Generative Adversarial Networks. [TPAMI]. *IEEE Transactions on Pattern Analysis and Machine Intelligence*. Advance online publication. DOI: 10.1109/TPAMI.2020.2970919 PMID: 32012000

Karras, T., Laine, S., Aittala, M., Hellsten, J., Lehtinen, J., & Aila, T. (2021). Alias-free generative adversarial networks. *Advances in Neural Information Processing Systems*, 34, 852–863. DOI: 10.48550/arXiv.2106.12423

Korshunov, P., & Marcel, S. (2018). Deepfakes: a new threat to face recognition? Assessment and detection. arXiv preprint arXiv:1812.08685. https://doi.org//arXiv.1812.08685DOI: 10.48550

Ledig, C., Theis, L., Huszar, F., Caballero, J., Cunningham, A., Acosta, A., & Shi, W. (2017). Photo-realistic single image super-resolution using a generative adversarial network. *Proceedings of the IEEE Conference on Computer Vision and Pattern Recognition*, 4681-4690. https://doi.org/DOI: 10.1109/CVPR.2017.19

Liu, M. Y., Breuel, T., & Kautz, J. (2017). Unsupervised image-to-image translation networks. Advances in Neural Information Processing Systems (NeurIPS), 700-710. https://doi.org/DOI: 10.48550/arXiv.1703.00848

Mirsky, Y., & Lee, W. (2021). The creation and detection of deepfakes: A survey. *ACM Computing Surveys*, 54(1), 1–41. DOI: 10.1145/3425780

Mirsky, Y., & Lee, W. (2021). The Creation and Detection of Deepfakes: A Survey. *ACM Computing Surveys*, 54(1), 1–41. DOI: 10.1145/3425780

Mirza, M., & Osindero, S. (2014). Conditional generative adversarial nets. arXiv preprint arXiv:1411.1784. https://doi.org//arXiv.1411.1784DOI: 10.48550

Miyato, T., Kataoka, T., Koyama, M., & Yoshida, Y. (2018). Spectral normalization for generative adversarial networks. arXiv preprint arXiv:1802.05957. https://doi.org//arXiv.1802.05957DOI: 10.48550

Miyato, T., Kataoka, T., Koyama, M., & Yoshida, Y. (2018). Spectral normalization for generative adversarial networks. *International Conference on Learning Representations (ICLR)*. https://doi.org/DOI: 10.48550/arXiv.1802.05957

Pidhorskyi, S., Adjeroh, D. A., & Doretto, G. (2020). Adversarial latent autoencoders. *Proceedings of the IEEE/CVF Conference on Computer Vision and Pattern Recognition (CVPR)*, 14104-14113. https://doi.org/DOI: 10.1109/CVPR42600.2020.01412

Radford, A., Metz, L., & Chintala, S. (2015). Unsupervised Representation Learning with Deep Convolutional Generative Adversarial Networks. *International Conference on Learning Representations (ICLR)*. https://doi.org/DOI: 10.48550/arXiv.1511.06434

Radford, A., Metz, L., & Chintala, S. (2016). Unsupervised representation learning with deep convolutional generative adversarial networks. arXiv preprint arXiv:1511.06434. https://doi.org//arXiv.1511.06434DOI: 10.48550

Rossler, A., Cozzolino, D., Verdoliva, L., Riess, C., Thies, J., & Nießner, M. (2019). FaceForensics++: Learning to detect manipulated facial images. arXiv preprint arXiv:1901.08971. https://doi.org//arXiv.1901.08971DOI: 10.1109/ICCV.2019.00009

Sauer, A., Schwarz, K., & Geiger, A. (2021). StyleGAN-XL: Scaling StyleGAN to large diverse datasets. Advances in Neural Information Processing Systems (NeurIPS). https://doi.org/DOI: 10.48550/arXiv.2202.00273

Tolosana, R., Vera-Rodriguez, R., Fierrez, J., Morales, A., & Ortega-Garcia, J. (2020). Deepfakes and beyond: A survey of face manipulation and fake detection. *Information Fusion*, 64, 131–148. DOI: 10.1016/j.inffus.2020.06.014

Tolosana, R., Vera-Rodriguez, R., Fierrez, J., Morales, A., & Ortega-Garcia, J. (2020). Deepfakes and Beyond: A Survey of Face Manipulation and Fake Detection. *Information Fusion*, 64, 131–148. DOI: 10.1016/j.inffus.2020.06.014

Verdoliva, L. (2020). Media forensics and deepfake detection: Theories, methods, and challenges. *IEEE Signal Processing Magazine*, 37(3), 42–50. DOI: 10.1109/MSP.2020.2978432

Verdoliva, L. (2020). Media Forensics and Deepfake Detection: Theories, Methods, and Challenges. *IEEE Signal Processing Magazine*, 37(3), 42–50. DOI: 10.1109/MSP.2020.2978432

Wang, T. C., Liu, M. Y., Zhu, J. Y., Tao, A., Kautz, J., & Catanzaro, B. (2018). High-resolution image synthesis and semantic manipulation with conditional GANs. *Proceedings of the IEEE Conference on Computer Vision and Pattern Recognition*, 8798-8807. https://doi.org/DOI: 10.1109/CVPR.2018.00917

Yu, J., Lin, Z., Yang, J., Shen, X., Lu, X., & Huang, T. S. (2018). Generative image inpainting with contextual attention. *Proceedings of the IEEE Conference on Computer Vision and Pattern Recognition (CVPR)*, 5505-5514. https://doi.org/DOI: 10.1109/CVPR.2018.00577

Zhang, H., Goodfellow, I., Metaxas, D., & Odena, A. (2018). Self-attention generative adversarial networks. *International Conference on Machine Learning*, 7354-7363. https://doi.org/DOI: 10.48550/arXiv.1802.05751

Zhang, H., Goodfellow, I., Metaxas, D., & Odena, A. (2019). Self-attention generative adversarial networks. *Proceedings of the 36th International Conference on Machine Learning*, 7354-7363. https://doi.org/DOI: 10.48550/arXiv.1805.08318

Zhu, J. Y., Park, T., Isola, P., & Efros, A. A. (2017). Unpaired image-to-image translation using cycle-consistent adversarial networks. *IEEE International Conference on Computer Vision*, 2223-2232. https://doi.org/DOI: 10.1109/ICCV.2017.244

Chapter 5
Advancements in Real-World Applications of Generative Adversarial Networks (GANs)

Keshika Jangde
https://orcid.org/0009-0003-4492-2461
Shri Shankaracharya Institute of Professional Management and Technology, Raipur, India

Ranu Pandey
Shri Rawatpura Sarkar University, Raipur, India

ABSTRACT

Generative Adversarial Networks (GANs) have been applied to various fields such as drug discovery, media detection, fashion, autonomous cars, healthcare, natural language processing, and engineering design. To improve unsupervised learning, GANs are made up of a generator network and discriminator that engage in adversarial training. In order to improve the fidelity of data distribution, the generator creates synthetic data that makes it difficult for the discriminator to distinguish between generated and real data. New developments in GAN architectures, such as DCGAN and WGAN, customise models for particular applications, such as hierarchical learning and stable distribution assessment. GANs impact is highlighted by their applications in drug discovery, image synthesis, and style transfer. They are transforming AI in the gaming, cybersecurity, and medical sectors by generating realistic data and encouraging innovation.

DOI: 10.4018/979-8-3693-7152-7.ch005

1. INTRODUCTION

Generative adversarial network (GAN) is an unsupervised learning neural network with an adversarial training setup. One Attack: Using GANs for Realistic Content Creation GANs or Generative Adversarial Networks are a radical AI technique capable of creating realistic text, images and videos. They employ two neural networks and a unique adversarial training process.

GANs have taken the world by storm and are used in many fields to generate high-quality and realistic data. Below are some Transformative Applications:

- **Image Synthesis**: GANs can produce realistic-looking images, such as improving grainy photos or producing realistic faces.
- **Style Transfer**: They have the ability to transform commonplace photos into artistic styles, like transforming a photograph into the look of well-known painters.
- **Text-to-Image Conversion**: Design and advertising benefit from the ability of GANs to convert descriptive text into realistic pictures.
- **Medical Imaging**: GANs are used in healthcare to improve diagnosis and treatment planning by producing thorough medical scans.
- **Entertainment**: They are employed in fashion design, deepfake videos, and virtual characters for computer games and films.

In deep learning, generative adversarial networks (GANs) have become a game-changer for producing diverse and high-quality data. Generative adversarial networks (GANs) have emerged as a transformative deep learning method that can produce high-quality, diverse data. In a GAN, a generator network generates data, and a discriminator network judges if it is real. An adversarial process allows the discriminator to learn features that help it distinguish between real and fake data, while the generator aims to create data that resembles real data (Chakraborty, Tanujit, et al.,2024).

Deep learning is yielding some of the best models such as generative adversarial network (GAN) architecture which shows a lot of potential in areas like language translation, image and video analysis, drug discovery, etc. In this regard, GANs can offer innovative solutions that have shown potential in drug research and development. In many areas of artificial intelligence and machine learning, their use is still growing (Tripathi, Satvik, et al., 2022).

The application of GANs in ophthalmic image processing seems promising when compared to traditional deep learning methods, however clinical validation is still in its early stages. The analysis of ocular pictures can be significantly enhanced by removing technological barriers and selecting suitable GAN techniques, opening up possibilities for deeper learning research using ophthalmology datasets (You, Aram, et al., 2022).

1.1. Types of GAN

Generative Adversarial Networks have developed into a number of varieties, each suited to a particular application and advancement in generative modelling. Some well-known GAN varieties are as follows (Porkodi, S. P., et al.,2023):

Figure 1. Generative adversarial network along with its different types of GAN models.

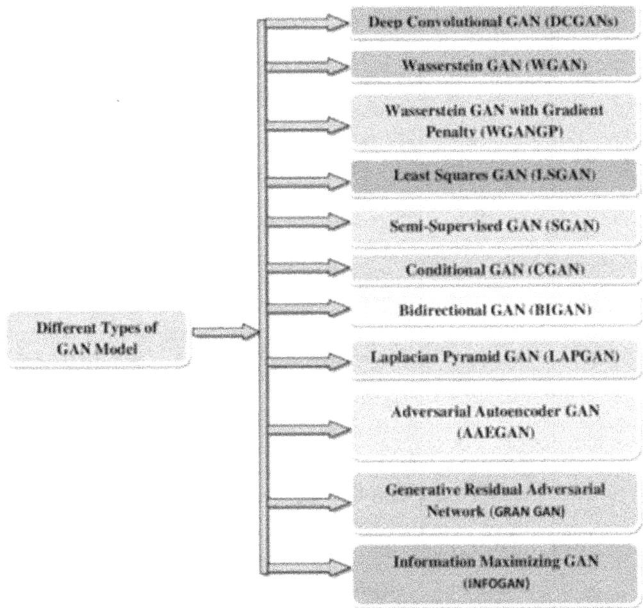

Table 1. Presents a comparative analysis of multiple GAN models using different criteria.

Types of GAN	Learning	Architecture	Objective methodology	Gradient updates	Performance comparison metrics	Application
GAN	Supervised	Multilayer perceptron	D's value function is growing while G's is decreasing.	In SGD, there are k steps for D and 1 step for G.	Log-likelihood	Medical Image Synthesis, Fashion and Design and Satellite Image Analysis
DCGAN	Unsupervised learning	CN with constraints	Discover the sequence of representations in G and D, starting from segments of things and ending with scenes.	SGD for both G and D using the Adam optimizer	Rates of accuracy and error	Image creation and synthesis
WGAN	Unsupervised learning	WGAN critic or the GAN discriminator	Calculating the disparity between the generated and actual data distributions	Earth Mover's range	Boost the learning process' stability	Image generation, image synthesis
WGANGP	Unsupervised learning	Lipschitz constraint	As an alternative to trimming weights, use the input's critic function.	critic's role	Enhancement of the WGAN	Creation and synthesis of images
LSGAN	Supervised & Semi supervised learning	Probably approximate correct (PAC)-style theorem	The resulting photos did not exhibit any noticeable degradation or mode collapse upon the removal of the batch normalizing layers.	non-vanishing gradient and cone-shaped	To achieve optimal performance in picture production	Image generation, image synthesis
SGAN	Semi supervised learning	Constraint based and CNN	Reduce the generator's training time and improve the quality of the generated samples.	basic classifier without generating element	Log-likelihood	Translation from image to image and style transfer

continued on following page

Table 1. Continued

Types of GAN	Learning	Architecture	Objective methodology	Gradient updates	Performance comparison metrics	Application
CGAN	Supervised learning	Multilayer perceptron	Reduce value function for G and raise value function for D contingent on more information	In SGD, there are k steps for D and 1 step for G.	Log-likelihood	Generation of conditional images
BIGAN	Supervised & Unsupervised learning	Deep multilayer neural networks	Recognize traits for related semantic tasks and use them in uncontrolled scenarios	------	Accuracy	Extensive image creation and synthesis
LAPGAN	Supervised learning	Laplacian pyramid of CN	The coarse-to-fine scale is used to generate the fashion picture.	-------	Estimation by humans and loglikelihood	Creating high-resolution images
AAE	Supervised, Semi supervised & Unsupervised learning	Autoencoders	Making an inference with the autoencoder's hidden code vector's posterior and prior distributions	SGD includes a step for regularization and reconstruction	Log-likelihood and rate of error	Generating data and learning representations
GRAN	Supervised learning	Recurrent CN with constraints	Images are created by making gradual changes to a "canvas."	G and D are updates for SGD.	Proposed is a generative adversarial metric.	Text Generation ,Music Generation, Video Generation& Speech Synthesis
INFOGAN	Unsupervised learning	Multilayer perceptron	Maximize mutual information in order to learn disentangled representations.	Both G and D are upgrades to SGD.	Learning through information metrics and representation	Image Editing, Data Augmentation & Semantic Data Generation

In this [Table 1] the different types of Generative Adversarial Networks (GAN) with architecture name are demonstrated.

Deep Convolutional Generative Adversarial Networks (DC-GAN) are used for medical image classification in order to enrich data on retinal scans from the APTOS Blindness dataset. The detection of diabetic retinopathy is accomplished with 98.66% accuracy using a ResNet50 model. The study shows how well DC-GAN works to improve classification performance and address class imbalance (Devi, Y. Sravani, et al.,2022). A technique to deal with data imbalance in aquatic ecosystem health (AEH) grade prediction that makes use of Wasserstein Generative Adversarial Networks (WGAN). Machine learning model performance is enhanced by the added data, as evidenced by F1-scores exceeding 0.9 in AEH grade predictions. With the possibility for additional improvement to take extreme occurrences in water management into consideration, this method improves prediction accuracy while reducing data collection efforts (Lee, Seoro, et al., 2021).

In GPR data interpretation, deep learning models have demonstrated success; nevertheless, it is difficult to obtain enough labelled data. This paper addresses the lack of data by proposing an enhanced least square generative adversarial network (LSGAN) model to produce high-precision GPR images. The LSGAN model performs better than existing GANs, exhibiting better adaptability for data augmentation and lower FID scores. GPR detection accuracy is increased by 10% when LSGAN-generated images are used for training. Future research will concentrate on improving picture variety using FDTD simulation and field data, as well as producing GPR images for additional subsurface objects (Yue, Yunpeng, et al.,2021). The ability of Generative Adversarial Networks (GANs) to produce realistic images has drawn a lot of interest in the field of medical image fusion. It outlines the fundamental GAN concepts, with focus on the model structure and training procedure. This touches on various GAN variants according to neural network architecture, network structures, and probability distribution. In this article, we explore the advantages and applications of GANs in the context of medical picture fusion, highlighting the ways in which they can improve the integration and quality of medical images. It covers the issues GANs face in this respect as well as potential paths for future work to make them function better (Zhou, Tao, et al,2023).

A denoising approach for low-dose chest imaging using CGANs - conditional generative adversarial networks - has been proposed. The approach trains on lung CT challenge and image datasets, effectively noise reduction with high-quality image restoration. When compared to the traditional approaches non-local means (NLM) and total-variation minimisation (TV), the CGAN method retains diagnostic features while offering superior image quality. For this reason, the present methodology enhanced the Structural Similarity Index Measure (SSIM) between 1.5 and 2.5 times in contrast to conventional ones. This denoised approach improves medical diagnosis with reduced radiation exposure and improved image quality (Kim, Hee-Joung, and Donghoon Lee.,2020). Data with anomalies is generally not balanced and also without labels; therefore, it is hard to identify an anomaly. Thus, various machine learning techniques exist for supervised, unsupervised, and semi-supervised learning. For one-class anomaly detection tasks such as network intrusion detection, we present two training schemes for BiGAN that decouple the discriminator from the generator. In anomaly detection, our proposed approaches enhance BiGAN's performance by attuning the training framework. These results exhibit tremendously increased efficacy in BiGAN by pre-training the generator and updating it independently from the discriminator. Additionally, future efforts could be directed toward multi-class anomaly detection for better classification (Kaplan, M. Oguz, et al. 2020).

This study examines several image synthesis techniques, including image-to-image and text-to-image translation, and it also explores potential avenues for further study. Through the development of a GAN model that can produce high-quality, editable images, the research laid the groundwork for upcoming developments in e-commerce, gaming, and art (Lamba, S. H., et al., 2023).

This is a new unsupervised clustering technique that combines density peak-spectral clustering (Dpeak-SC) and adversarial autoencoder (AAE) for unlabelled ISAR images of space targets. After features are extracted from radar image data using AAE, clustering is done using spectral and density peak techniques. Self-supervised training, feature extraction in low-dimensional space, and decision graph creation for cluster identification are all part of the approach. The method's ability to accurately cluster unlabelled ISAR picture data is demonstrated by simulation results (Yang, Hong, et al. 2022). The presented research proposes a Gated Graph Neural Network-based Graph Recurrent Attention Network (GGNN-GRAN) to handle the problem of pedestrian direction recognition in complex traffic scenarios. GRAN is used to mine dependencies between body parts when a body parts graph (head, arms, and legs) is constructed on feature maps. Graph node associations are dynamically represented during training via an attention-based adjacency matrix. With classification rates of 97%, 91%, and 90%, respectively, experiments on three benchmarks (PDC, PDRD, and Cityscapes) demonstrate that the system outperforms the state-of-the-art techniques now in use (Li, Xiao, et al.,2023).

This paper discusses the problems faced by Facial Sketched-Real Image Retrieval (FSRIR) within Content-Based Image Retrieval (CBIR) systems, which aim to find images similar by means of visual features. The authors produced a new dataset called Extended Sketched-Real Image Retrieval (ESRIR) with 53,000 facial sketches versus 53,000 real images expropriated from the Chinese University Face Sketch (CUFS) dataset. The paper introduces three new retrieval techniques based on the Vision Transformer (ViT), InfoGAN, and convolutional autoencoder in an unsupervised setting. These additions are meant to enhance facial recognition systems' content-based similarity matching and retrieval (Sabry, Eman S., et al.,2023).

1.2 Architecture of GANs

The Generator and the Discriminator are the two primary parts of a Generative Adversarial Network (GAN).

Generator: The purpose of the generator is to produce synthetic data that closely mimics real data, including text, graphics, and so forth. Starting with random data, it gradually improves it in response to discriminator criticism. The generator aims to create artificial data that is so realistic that the discriminator is unable to distinguish it from actual data. Produces artificial data, much like an artist attempting to create lifelike artwork.
Discriminator: Acts as a critic, isolates authentic data from fraudulent data. Making the distinction between authentic and fraudulent data is the discriminator's job. Whether the data it gets is authentic (from the training dataset) or fraudulent (generated by the generator) is assessed. It gives the generator feedback so that it can provide better results.

Figure 2. Basic architecture of GAN

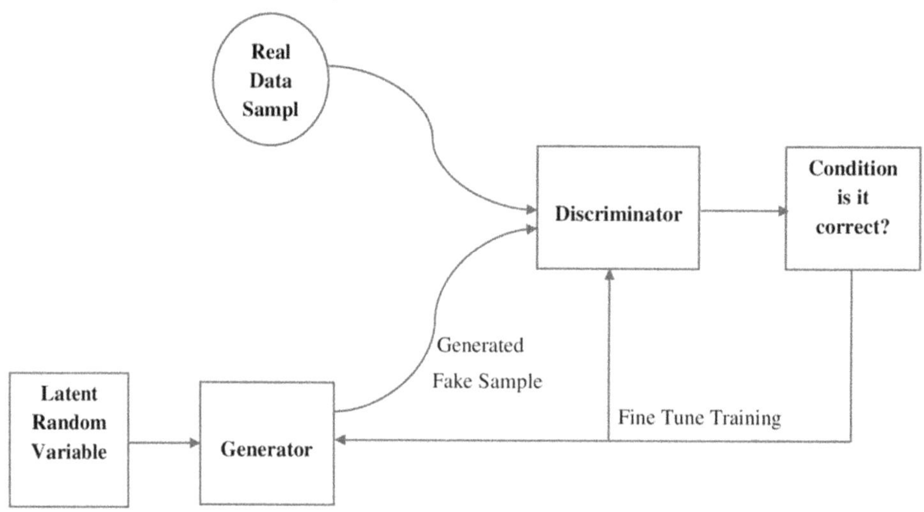

Working Principle of GAN

1. **Initialization:** A discriminator (D) and a generator (G) neural network are constructed. G's job is to create new data that closely resembles genuine data, such as text or photos. D takes on the role of a critic, attempting to

discern between the data produced by G and actual data (from a training dataset).

2. **Generator's First Move:** G accepts a vector of random noise as input. G's creation process begins with this noise vector, which has random values in it. The noise vector is converted into a new data sample, such as a produced image, by G using its internal layers and learnt patterns.
3. **Discriminator's Turn:** Two types of inputs are given to D:
    - Authentic data samples taken from the training set.
    - The data samples produced in the preceding stage by G. D's task is to examine each input and ascertain whether it is authentic data or a fabrication by G. It produces a likelihood score that ranges from 0 to 1. A score of 1 means the data is probably authentic, whereas a score of 0 means it is not.
4. **The Learning Process:** Now, the adversarial part comes in
    - In the event that D properly recognises created data as false (score near 0) and actual data as real (scoring near 1), both G and D receive a little reward. That's because they're both performing their jobs effectively.
    - The secret, though, is to keep getting better. It won't pick up any knowledge if D continually recognises everything. Therefore, G is supposed to finally deceive D.
5. Generator's Improvement:
    - G is on the right track when D incorrectly classifies G's creation as real (scoring around 1). In this instance, D gets penalised for being duped, whereas G receives a noteworthy positive update.
    - G is able to produce more realistic data by refining its creation method thanks to this feedback.
6. Discriminator's Adaptation:
    - The discrimination skills of D are further enhanced if, on the other hand, it properly detects G's phoney data (score near 0), but G is not rewarded.
    - Both networks are gradually improved by this continuous competition between G and D.

G becomes more adept at producing realistic data as training goes on, making it more difficult for D to distinguish between the two. Ideally, G improves to the point where D is unable to consistently discern between authentic and fraudulent data. G is now regarded as well-trained and capable of producing fresh, accurate data samples.

1.3 Diverse Applications of Generative Adversarial Networks (GANs)

There are plenty of applications across different disciplines as a result of Generative Adversarial Networks' (GANs) capacity to produce new data that closely mimics actual data distributions. These are some of the most recent real-world applications of GANs.

1. **Image and Video Generation:** Generating realistic images and videos is a common application of GANs. New developments have raised both the quality and variety of content produced, which is beneficial for industries like entertainment, fashion, and design. An innovative architecture for creating videos that improves spatiotemporal consistency without the need for complex 3D models. Through information sharing between neighbouring frames, it enhances both high-level structures and low-level details. It introduces the S3 score for quantitative evaluation via downstream tasks and achieves state-of-the-art outcomes on UCF-101 with Inception Score. The TSB architecture's effectiveness in increasing video production is established by validation on multiple datasets, such as MaisToy, which shows improved performance over BigGAN, backed by solid qualitative evidence (Munoz, Andres, et al.,2021).
2. **Data Augmentation:** Machine learning models use augmented reality (GANs) to improve training data. GANs contribute to the enhancement of model robustness and generalization by producing synthetic data that closely mimics real data. The work suggests DAG, a framework designed for GAN training with augmented data, in order to overcome the problem of costly data collecting in fields like medicine. In order to align augmented data with original distributions, it proposes DAG and argues against classical data augmentation, which may lead to misleading GANs. The efficiency of DAG in lowering JS divergence is confirmed by theoretical study. Experiments using a variety of GAN models and datasets (natural and medical images) show that DAG consistently improves both generator and discriminator learning, leading to notable performance gains (Tran, Ngoc-Trung, et al.,2021).
3. **Medical Image Analysis**: GANs improve the quality of medical images for precise diagnosis and therapy by detecting anomalies, segmenting images, and augmenting them. By providing quick, high-quality artificial image production, they tackle issues like exorbitant expenses, radiation exposure, and data gathering time. GANs do exceptionally well in image-to-image translation, which includes translating between various MRI strengths to produce multimodal datasets and improving MRI from CT images (Suganthi, K., 2021).

4. **Video Synthesis and Prediction:** Video compression, action recognition, and autonomous driving all use GANs to create and forecast future video frames. A more sophisticated technique uses non-rigid deformation and affine modifications to distinguish moving objects from the backdrop for improved frame prediction. In terms of accuracy and visual quality, this technology performs better than current techniques while reducing problems like distortion and ripping. When tested on the Cityscapes and KITTI datasets, it performs better than expected (Wu, Yue, et al.,2020).
5. **Text-to-Image Synthesis:** A common use in the creative industries, e-commerce (product visualization), and content development is the generation of images from written descriptions using GANs.

Text-to-image synthesis utilizes both more current unsupervised deep learning (DL) techniques, specifically deep generative models like GANs and DCNNs, and more traditional supervised techniques based on word-image correlations to translate written descriptions into images. The automation of image synthesis from natural language has garnered considerable interest from a variety of industries, including virtual reality, gaming, and design. A taxonomy classifying GAN-based text-to-image synthesis into semantic enhancement, resolution enhancement, diversity enhancement, and motion enhancement GANs is proposed in the survey, which also evaluates state-of-the-art models and describes architectures like GANs and DCNNs. It examines their development, uses in producing a variety of content, including interiors, birds, and human faces, and talks about difficulties and potential future paths for the discipline (Agnese, Jorge, et al.,2020).

6. **Super-Resolution Imaging:** In domains including satellite imaging, microscopy, and surveillance, GANs can improve image quality and resolution.

Image compression, remote sensing, medical imaging, and other industries all depend on high-resolution images. Reconstruction with super-resolution seeks to use particular techniques to transform low-resolution images into high-resolution ones. Though thorough studies incorporating their innovations are scarce, Generative Adversarial Networks (GANs) have transformed this discipline (Wang, Xuan, et al.,2023).

7. **Art and Creativity:** The use of these machines to create new structures and paintings demonstrates the potential of GANs in the fields of creativity.

Pablo Picasso once said, "art is the lie that we reveal the truth." This statement shows the commitment by humans to art throughout the centuries. As the years go by, there are efforts from scientists to understand the possibilities of computer generated art, especially under deep learning and generative adversarial networks (GANs). It is possible to create realistic human faces along with captions from images with GANs. Recent progress in creative power of visual arts, music, and literature produced with the help of GANs is the focus of this overview (Shahriar and Sakib., 2022).

8. **Anomaly Detection:** Anomaly detection using GANs relies on understanding the normal data distribution and then pinpointing any anomalies which is advantageous for cybersecurity, fraud detection, and quality control. With the help of GANs, it's easier to differentiate between normal and abnormal patterns in data which helps in fraud prevention, medical diagnosis, and cybersecurity. GANs help fill the gap by synthesizing new data to help eliminate insufficient abnormal data which has plagued most traditional anomaly detection methods. This paper explores the application of different designs of GANs such as DCGANs and cGANs in intrusion detection, monitoring, and diagnosis. Also, it builds the foundation for designing efficient GANs based anomaly detection systems and ideas for further research (Sabuhi, Mikael, et al., 2021).
9. **Drug Discovery and Molecular Design:** GAN are employed to generate distinctive compounds with specified properties, which speeds up the process of developing new medications and materials.Recent studies demonstrate how deep learning algorithms and artificial intelligence, in particular, will play a critical role in changing medication discovery and design. This review concentrates on these domains' use of generative deep learning techniques, such as GAN frameworks. Peptide/protein synthesis, dimension reduction of single-cell data in preclinical phases, and molecular de novo design are among the applications it investigates. In addition, the shortcomings of earlier research employing GANs for drug development and ends with recommendations for future directions and obstacles in the field (Tripathi, Satvik, et al.,2022).
10. **Deepfake Detection:** GANs are being used in deeperfake technology to create increasingly sophisticated techniques for identifying edited or synthetic material. These methods assist in spotting deepfake pictures, sounds, and videos in order to dispel false information and guard against nefarious usage.

AI, machine learning, and deep learning developments over the past few decades have transformed multimedia manipulation and produced tools for a wide range of acceptable applications, including teaching and entertainment. These technologies have been abused, though, and sometimes for nefarious purposes. The incredibly realistic false multimedia produced by malicious actors, or “Deepfakes,” is intended to propagate misinformation, provoke political unrest, incite hatred, and even harass or extort people. These artificially produced images, with their remarkable realism and quality, emphasize a growing issue in today's digital environment by demonstrating the dual potential of technology advancement for both positive and negative effects (Rana, Md Shohel, et al.,2022).

11. **Game Development:** GANs can generate locations, characters, and textures, which is useful for game designers and procedural content producers. Applications for Generative Adversarial Networks (GANs) include classification, super-resolution, image synthesis, style transfer, and semantic editing (Alqahtani, Hamed, et al.,2021).

Figure 3. Applications of generative adversarial networks GAN

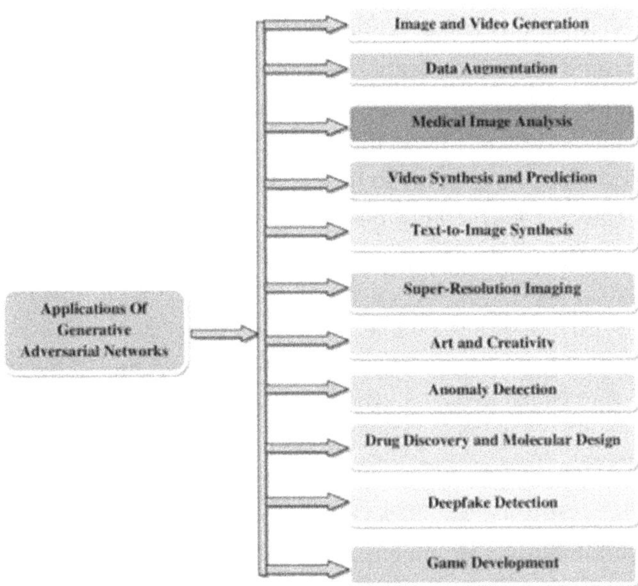

The different kinds of applications of GAN are illustrated in the above [figure 3].

The application of Generative Adversarial Networks (GAN) technology facilitates the fabrication of DeepFake content which has greatly increased the accessibility of face-swapping in images and videos which cripples the line between real and fake. This proliferation of unverified information has become an epidemic. The research digs into the performance of GAN-based models for DeepFake detection, especially for those based on Deep Convolution GANS. With the proposed model, higher accuracy in detecting fake content is achieved irrespective of the small sizes of data sets employed. In light of the rising concern of "fake news", the research clearly points out the importance of developing GAN detection techniques and focuses on improving the generalization of the model and the resolution of mode collapse problems (Kumar, Manoj, and Hitesh Kumar Sharma., 2023).

Generative Adversarial Networks (GANs) swapped faces in images and videos, heavily using its design to create DeepFake content, which produces a sudden challenge for verifying the authenticity of media. Malicious and misinformative content has become a socially contagious issue. In particular, the study examines the efficacy of GAN -based models, specifically Deep Convolution–based GAN, to identify DeepFakes. Our proposed model is traced over very disjointed data where the much more substantial series of iteration to achieve the discovery to fraudulent value of media even with small dataset is significantly improved. The paper also calls for improved GAN detection techniques for future challenges of fake media and points to challenges such as mode collapse and improving model generalisation (Wesselkamp, Vera, et al.,2022). While passive detection methods address the issue of DeepFakes after their generation, significant security concerns are raised by approach, and proactive measures such as perturbations are easily bypassed. In this work we exploit a reconstructive regularisation of the loss function of the GAN, in order to embed visible watermarks in DeepFakes through a proactive GAN-based approach. The watermarks, readable both by automatic and by human systems, serve as a warning and are used to detect malicious DeepFakes (Nadimpalli, et al.,2024).

Deepfake detection necessitates creative solutions because current techniques are inexplicable and non-generalizable. This study suggests the CTF method, which uses GAN Specific Frequencies (GSF) as distinct fingerprints found using Discrete Cosine Transform (DCT) to recognise content produced by GANs. The approach provides useful forensic insights and is quick, understandable, resistant to different types of attacks, and generalisable across datasets. It is further improved by a G-boost classifier (Giudice, Oliver, et al., 2021). AI and computer vision developments allow machines to imitate human characteristics. For example, GANs, which creatively rearrange pixels, can produce realistic-looking false images. Notwithstanding its advantages, this technology presents problems since GAN-generated images, which frequently circulate on social media, feed fraud, privacy concerns, and false information. In order to differentiate between real and fake in a minimax game, GANs use a discriminator network in conjunction with a generator network to generate images. To prevent false information and safeguard privacy, researchers have created techniques to identify phoney photos. This chapter examines the most recent GAN detection methods and points up areas that need more study (Arora, Tanvi, and Rituraj Soni., 2021).

2. CONCLUSION

Generative Adversarial Networks (GANs) are a new way to do unsupervised learning, using two parts called a discriminator and a generator that train against each other. GANs improve how models can create data that looks very real. They are powerful in many areas like finding anomalies, analyzing medical images, and creating deepfakes. In medicine, GANs help make more data, improve image quality, and get better predictions, like in spotting diabetic retinopathy and guessing environmental health ratings. In other fields, GANs create detailed images, help look for new drugs, and enhance the quality of chest and GPR images. However, the ability of deepfakes to create misleading media is a big problem. As a result, the need for reliable and widely usable methods to stop the spread of false information and protect privacy is increasing. Several new techniques have been introduced to spot GAN-generated content, such as using GAN Specific Frequencies (GSF) for deepfake detection, which provides a clear and effective solution. While GANs have many groundbreaking uses, more research is needed to enhance their capabilities and address the ethical issues their misuse creates. GAN detection and image creation methods hold a lot of promise for innovation and safeguarding data integrity.

REFERENCES

Agnese, J., Herrera, J., Tao, H., & Zhu, X. (2020). A survey and taxonomy of adversarial neural networks for text-to-image synthesis. *Wiley Interdisciplinary Reviews. Data Mining and Knowledge Discovery*, *10*(4), e1345. DOI: 10.1002/widm.1345

Alqahtani, H., Kavakli-Thorne, M., & Kumar, G. (2021). Applications of generative adversarial networks (gans): An updated review. *Archives of Computational Methods in Engineering*, *28*(2), 525–552. DOI: 10.1007/s11831-019-09388-y

Arora, T., & Soni, R. (2021). A review of techniques to detect the GAN-generated fake images. *Generative Adversarial Networks for Image-to-Image Translation*, 125-159.

Chakraborty, T., Reddy K S, U., Naik, S. M., Panja, M., & Manvitha, B. (2024). Ten years of generative adversarial nets (GANs): A survey of the state-of-the-art. *Machine Learning: Science and Technology*, *5*(1), 011001. DOI: 10.1088/2632-2153/ad1f77

Devi, Y. S., & Kumar, S. P. (2022). DR-DCGAN: A deep convolutional generative adversarial network (DC-GAN) for diabetic retinopathy image synthesis. *Webology (ISSN: 1735-188X)*, *19*(2).

Giudice, O., Guarnera, L., & Battiato, S. (2021). Fighting deepfakes by detecting gan dct anomalies. *Journal of Imaging*, *7*(8), 128. DOI: 10.3390/jimaging7080128 PMID: 34460764

Kaplan, M. O., & Alptekin, S. E. (2020). An improved BiGAN based approach for anomaly detection. *Procedia Computer Science*, *176*, 185–194. DOI: 10.1016/j.procs.2020.08.020

Kim, H. J., & Lee, D. (2020). Image denoising with conditional generative adversarial networks (CGAN) in low dose chest images. *Nuclear Instruments & Methods in Physics Research. Section A, Accelerators, Spectrometers, Detectors and Associated Equipment*, *954*, 161914. DOI: 10.1016/j.nima.2019.02.041

Kumar, M., & Sharma, H. K. (2023). A GAN-based model of deepfake detection in social media. *Procedia Computer Science*, *218*, 2153–2162. DOI: 10.1016/j.procs.2023.01.191

Lamba, S. H., Singh, R. K., & Soni, E. K. (2023). Generative adversarial network (GAN) to generate realistic images. *International Journal for Research in Applied Science and Engineering Technology*, *11*(4), 2190–2196. DOI: 10.22214/ijraset.2023.50306

Lee, S., Kim, J., Lee, G., Hong, J., Bae, J. H., & Lim, K. J. (2021). Prediction of aquatic ecosystem health indices through machine learning models using the WGAN-based data augmentation method. *Sustainability (Basel)*, *13*(18), 10435. DOI: 10.3390/su131810435

Li, X., Ma, S., Shan, L., Liu, S., Chai, S., & Li, X. (2023). GRAN: Graph recurrent attention network for pedestrian orientation classification. *Complex & Intelligent Systems*, *9*(1), 891–908. DOI: 10.1007/s40747-022-00836-0

Munoz, A., Zolfaghari, M., Argus, M., & Brox, T. (2021). Temporal shift GAN for large scale video generation. In *Proceedings of the IEEE/CVF Winter Conference on Applications of Computer Vision* (pp. 3179-3188). DOI: 10.1109/WACV48630.2021.00322

Nadimpalli, A. V., & Rattani, A. (2024). Proactive deepfake detection using gan-based visible watermarking. *ACM Transactions on Multimedia Computing Communications and Applications*, *20*(11), 1–27. DOI: 10.1145/3625547

Porkodi, S. P., Sarada, V., Maik, V., & Gurushankar, K. (2023). Generic image application using gans (generative adversarial networks): A review. *Evolving Systems*, *14*(5), 903–917. DOI: 10.1007/s12530-022-09464-y

Rana, M. S., Nobi, M. N., Murali, B., & Sung, A. H. (2022). Deepfake detection: A systematic literature review. *IEEE Access : Practical Innovations, Open Solutions*, *10*, 25494–25513. DOI: 10.1109/ACCESS.2022.3154404

Sabry, E. S., Elagooz, S. S., Abd El-Samie, F. E., El-Shafai, W., El-Bahnasawy, N. A., El-Banby, G. M., & Ramadan, R. A. (2023). Image retrieval using convolutional autoencoder, infogan, and vision transformer unsupervised models. *IEEE Access : Practical Innovations, Open Solutions*, *11*, 20445–20477. DOI: 10.1109/ACCESS.2023.3241858

Sabuhi, M., Zhou, M., Bezemer, C. P., & Musilek, P. (2021). Applications of generative adversarial networks in anomaly detection: A systematic literature review. *IEEE Access : Practical Innovations, Open Solutions*, *9*, 161003–161029. DOI: 10.1109/ACCESS.2021.3131949

Shahriar, S. (2022). GAN computers generate arts? A survey on visual arts, music, and literary text generation using generative adversarial network. *Displays*, *73*, 102237. DOI: 10.1016/j.displa.2022.102237

Suganthi, K. (2021). Review of medical image synthesis using GAN techniques. In *ITM Web of Conferences* (Vol. 37, p. 01005). EDP Sciences.

Tran, N. T., Tran, V. H., Nguyen, N. B., Nguyen, T. K., & Cheung, N. M. (2021). On data augmentation for GAN training. *IEEE Transactions on Image Processing*, *30*, 1882–1897. DOI: 10.1109/TIP.2021.3049346 PMID: 33428571

Tripathi, S., Augustin, A. I., Dunlop, A., Sukumaran, R., Dheer, S., Zavalny, A., Haslam, O., Austin, T., Donchez, J., Tripathi, P. K., & Kim, E. (2022). Recent advances and application of generative adversarial networks in drug discovery, development, and targeting. *Artificial Intelligence in the Life Sciences*, *2*, 100045. DOI: 10.1016/j.ailsci.2022.100045

Wang, X., Sun, L., Chehri, A., & Song, Y. (2023). A review of GAN-based super-resolution reconstruction for optical remote sensing images. *Remote Sensing (Basel)*, *15*(20), 5062. DOI: 10.3390/rs15205062

Wesselkamp, V., Rieck, K., Arp, D., & Quiring, E. (2022, May). Misleading deep-fake detection with gan fingerprints. In 2022 IEEE Security and Privacy Workshops (SPW) (pp. 59-65). IEEE.

Wu, Y., Gao, R., Park, J., & Chen, Q. (2020). Future video synthesis with object motion prediction. In *Proceedings of the IEEE/CVF Conference on Computer Vision and Pattern Recognition* (pp. 5539-5548). DOI: 10.1109/CVPR42600.2020.00558

Yang, H., Ding, W., & Yin, C. (2022). AAE-Dpeak-SC: A novel unsupervised clustering method for space target ISAR images based on adversarial autoencoder and density peak-spectral clustering. *Advances in Space Research*, *70*(5), 1472–1495. DOI: 10.1016/j.asr.2022.05.068

You, A., Kim, J. K., Ryu, I. H., & Yoo, T. K. (2022). Application of generative adversarial networks (GAN) for ophthalmology image domains: A survey. *Eye and Vision (London, England)*, *9*(1), 6. DOI: 10.1186/s40662-022-00277-3 PMID: 35109930

Yue, Y., Liu, H., Meng, X., Li, Y., & Du, Y. (2021). Generation of high-precision ground penetrating radar images using improved least square generative adversarial networks. *Remote Sensing (Basel)*, *13*(22), 4590. DOI: 10.3390/rs13224590

Zhou, T., Li, Q., Lu, H., Cheng, Q., & Zhang, X. (2023). GAN review: Models and medical image fusion applications. *Information Fusion*, *91*, 134–148. DOI: 10.1016/j.inffus.2022.10.017

Chapter 6
Image Generation with GANs

Rupali Vyas
https://orcid.org/0009-0006-1382-1141
Shri Shankaracharya Institute of Professional Management and Technology, Raipur, India

Rajendra Kumar Pandey
Shri Shankaracharya Institute of Professional Management and Technology, Raipur, India

ABSTRACT

This chapter explores the functionalities of Generative Adversarial Networks (GANs), which are employed for the purpose of image synthesis, surpassing conventional methods in their advancements. GANs use a generator and discriminator network for increasing image quality through adverse training. This introduction discusses the progression of methods used for image generation before the advent of GANs. Additionally, it provides an overview of the fundamental principles behind GAN training. While more sophisticated GAN architectures, such as DCGANs, utilize CNN to ensure stability, StyleGAN improves the accuracy of synthesis. Conditional GANs and impact of having domain specific using SRGAN, Text-to-Image to help boost the image quality undergone at major phases. Datasets and pre-trained models such as ImageNet, CelebA, CIFAR correspondingly in significant for training data. The chapter also addresses the application in art, fashion and garment design, medicine for imaging to medical its usage educations media affects creativity and AI training.

DOI: 10.4018/979-8-3693-7152-7.ch006

1. INTRODUCTION

Image production fulfills a significant need for original visual content, catering to large sectors. It aids the creative process of inquiry through the depiction of ideas and, the construction of variations thereof and conquers the limits either by filling in missing information or conjuring up new realities visually . It fosters better communication using product visualization and data storytelling. And ultimately, it economizes on efforts while automating content and prototype developments, giving it a helpful tool in various areas like entertainment, healthcare, or education.

The human impulse to produce art and images has been with us for thousands of years, from cave drawings to digital masterpieces, continuously pushing the limits of visual expression. Today, AI makes a major leap ahead with generative adversarial networks, a terrific tool for creating fresh, amazingly photorealistic images. This chapter digs into the fascinating world of GANs, uncovering the core concepts that enable them to produce such innovative and breakthrough outputs. We'll begin on a trip to learn the secret sauce behind GANs(Goodfellow et al., 2014a). Now, we will witness the mesmerizing process of adversarial training: a "generator" network working as an artist to produce images that look increasingly real, and its competitor a painstakingly careful discriminator network who serves as the ultimate art critic, Excel separating the real from the fabricated. In ongoing struggle, development to both is assured, and so, the forefront in image generation is constantly pushed.(Gu et al., 2020; Tan et al., 2023).

This chapter will explore the tremendous potential of GANs, exhibiting their power to construct totally new visuals and achieve artistic feats like style transfer. We'll also dig into the problems and ethical implications surrounding this powerful technology. Finally, we'll peak into the future, where GANs hold great promise for transforming the way we interact with and produce visual content.

Think of being baffled as we release the artist within - the might of Generative Adversarial Networks in producing ultimately, entirely new visual realms. Image generation has come a long way from the rule-based systems in imitation of real-world tools to sophisticated deep approaches to learning as shown in **fig 1**. GANs are on the very frontier of this area. These brilliant models do entail two neural networks embroiled in an artistic duel:

Generator: The creative force, starting with random noise and shaping it into new images.

Discriminator: The discerning critic, evaluating both real images and the generator's creations to identify fakes.

Figure 1. Generator and Discriminator in a GAN The two basic components of GAN comprise a generator that creates data samples and a discriminator that checks the legitimacy of the produced samples.

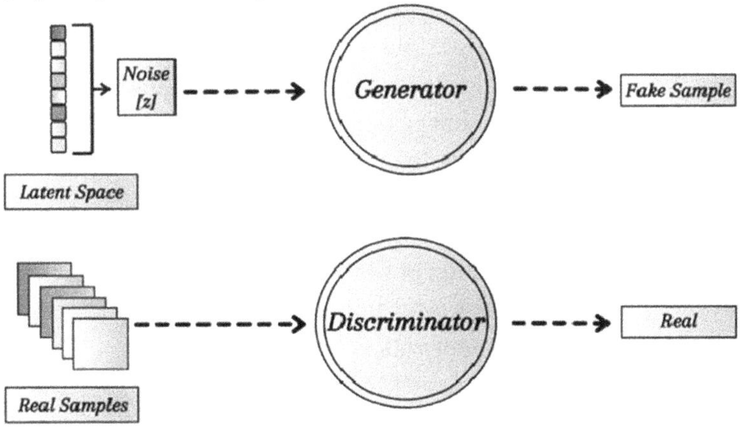

In GANs, the goal is to improve a generator's realism and deceive a discriminator to create synthetic data that looks real. V(D,G) for a discriminator D and a generator G:

$$\min_G \max_D \mathbf{V}(\mathbf{D},\mathbf{G}) = \mathbf{E}_{x \sim p_{dt}(x)}[\log D(x)] + E_{z \sim p_z(z)}[\log(1 - D(G(z)))] \quad (1)$$

Discriminator *D(x)*

- *D(x)* represents the probability that the input *x* is from the real data distribution $p_{data}(x)$
- The discriminator's role is to maximize its ability to classify real samples xx as real *(logD(x))* and synthetic samples *G(z)* as fake *(log(1−D(G(z))))*.

Generator G(z):

- The generator maps a latent variable *z*, sampled from a prior distribution $p_z(z)$ (e.g., Gaussian noise), to a synthetic sample *G(z)*.
- Its goal is to minimize *log(1−D(G(z)))*, which implies fooling the discriminator into classifying synthetic data as real.

Real Data Term $E_{x\ pdata}(x)[\log D(x)]$:

- This term encourages the discriminator to correctly classify samples drawn from the real data distribution *pdata*.

Synthetic Data Term $E_{z\sim pz(z)}[\log(1-D(G(z)))]$:

- This term ensures the discriminator penalizes the generator for producing samples *G(z)* that it identifies as synthetic.
- Simultaneously, the generator aims to produce *G(z)* such that *D(G(z))≈1*, effectively minimizing this term.

Through this ongoing competition, both networks improve. The generator learns to produce ever-more realistic images, while the discriminator hones its ability to spot forgeries. This results in the creation of stunningly realistic and creative new images. GANs hold immense potential for generating entirely new visuals, completing damaged photos, mimicking artistic styles, and even enhancing image resolution. However, challenges and ethical considerations remain. Training GANs is partly a balancing act, and the generated images should be fair and neutral. The possibility of making realistic deep fakes means their creation and usage must be done properly. In the future, researchers are focusing on making the visuals created by GANs more realistic with better control. This encompasses applications beyond photos, into the formation of 3D objects, the creation of films, and even text-to-image synthesis. As research proceeds, we will witness even more remarkable developments in viewing and producing visual media content with GANs.

1.1 Traditional vs. GAN-Based Image Generation: A Tale of Two Techniques

The world of image generation altered forever with the emergence of Generative Adversarial Networks, otherwise termed GANS. This means that old methods served us extremely well in many respects for many years; GANs offer a new approach defined by quite distinct advantages and constraints as shown in **table 1**. Read following, which will help you grasp the important differences:

1.1.1 Traditional Image Generation Techniques

Traditional image creation methods can be divided down into two primary categories: rule-based systems and statistical methods.

A. Rule-based methods work much like an electronic box of tools for the digital artist, where algorithms and tools are predefined to replicate the physical world with brushes of various textures and behaviors. This approach places the artist in a comfortable and controlled atmosphere wherein one may directly modify the parameters to acquire the desired effects. Furthermore, one has an understanding regarding what sort of reasoning resides behind these procedures, which makes them predictable and trustworthy for specific tasks. Another drawback of rule-based systems is that they are not inventive. Due to specified elements, they are bound to get repetitive and, therefore, are frequently ineffectual in developing entirely new acclaim or extremely inventive content.

B. On the other hand, statistical algorithms are trained on actual images. Imagine learning from a huge database of textures and backdrops what constitutes a realistic pattern. This goes quite a distance to create realistic textures or backgrounds. Statistical approaches can capture just a minimum perspective of the complete richness of natural images. Though able to produce realistic textures, these technologies quite often fail to be able to reproduce the complete range of nuances and variances that exist in reality.

1.1.2 Generative Adversarial Networks (GANs)

Deep learning, in this scenario, comes in with Generative Adversarial Networks, GANs, which will have two neural networks pitted against one other in a type of competitive dance. On the one hand, there is the generator, working like a creative artist. In essence, it is fed random noise, which it tries to transform into a realistic image by several adjustments. On the other side is its rival, dubbed a discriminator, who is a caustic art critic. It looks through both natural photographs and what the generator comes up with to be able to distinguish which one is phony. Because this happens repeatedly, both networks get better. The generator is going to be better at creating highly realistic forgeries, and the discriminator is getting better at being able to detect them more precisely. This approach has immense promise. GANs may create extraordinarily lifelike visuals that frequently blur the line between fiction and reality. It is, however, in sheer creativity that they excel: cranking out new and inventive content something which was almost impossible to produce with prior approaches. But all this power comes with challenges. Training GANs can be difficult and involves careful balancing between the two networks. Unlike traditional approaches with strict human control, GANs may be less predictable regarding exactly what material will be created. More than that, the level of realism that can be reached in photography today has already raised some ethical concerns since deepfakes carry hazards of misinformation and manipulation, as shown in **fig 2**.

Figure 2. Workflow of the Generative Adversarial Network: A formal depiction of how GANs train the process, including data production, assessment, and feedback loops between the generator and discriminator.

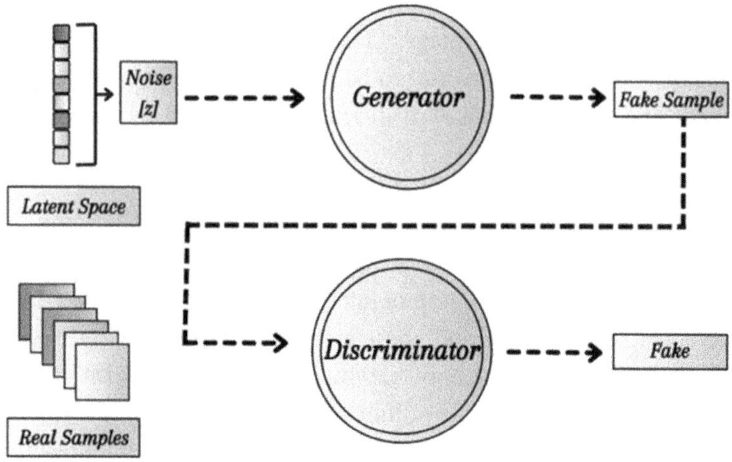

This **table 1** explores the key differences between traditional data generation methods and Generative Adversarial Networks (GANs). While both approaches have their strengths and weaknesses, GANs offer a revolutionary leap forward in terms of realism and creative potential.

Table 1. Traditional vs. deep learning approaches for data generation

Feature	Traditional Methods	Generative Adversarial Networks (GANs)
Approach	Rule-based systems, Statistical methods	Deep learning with two competing neural networks
Pros	- User Control - Interpretability	- Unmatched Realism - Unbridled Creativity
Cons	- Limited Creativity - Realistic Limitations	- Training Complexity - Limited Control - Ethical Considerations
Analogy	Artist using pre-defined brushes	Artist (Generator) creating based on critic's (Discriminator) feedback
Key Strength	Control and interpretability	Unmatched realism and creativity

The Adversarial Process: A Competitive Collaboration

The interplay between the generator and discriminator may be regarded as a high-stakes game of cat and mouse:

- The generator creates a batch of fake images.
- The discriminator examines the images along with actual ones.
- It is the discriminator that delivers the input required to discover what precisely made the images recognizable as phony.
- The feedback is utilized by the generator to develop its method. The process repeats, with both networks growing more complex.
- This back-and-forth resembles a master forger and investigator continually attempting to outwit one other. As the forger (generator) advances, the detective (discriminator) must grow more smart, and vice versa.

Choosing the Right Approach:

The best approach depends on your specific needs and priorities. Here's a quick guide:

1. Traditional Methods: Ideal for cases when you need tight control over the image or prioritize interpretability of the procedure.
2. GANs: Perfect for unleashing creativity and generating incredibly realistic visuals, especially when complete uniqueness is sought.

2. TYPES OF GAN

There are several types of Generative Adversarial Networks (GANs), each with its own strengths and applications. Here's a breakdown of some typical types:

2.1 Foundational GANs

2.1.1 Vanilla GAN

A Vanilla GAN is a two-part technique for synthesizing realistic data. One portion, the generator, takes random noise and molds it into something meaningful, like an image. The other portion, the discriminator, serves as a merciless critic, trying to separate actual data from the generator's concoctions. This antagonistic training pushes them against each other. The generator seeks to trick the discriminator with ever-more-realistic outputs, while the discriminator refines its analysis to catch the fakes. Through a continuous training loop with backpropagation, both models improve. The outcome? A generator capable of creating data that nearly mimics the actual thing. However, vanilla GANs can be unstable, providing the path for more complicated GAN architectures to expand upon this core principle(Goodfellow et al., 2014b) as shown in **fig 3**.

Figure 3. Components of a Vanilla GAN: The architecture of the Vanilla GAN: a basic generator model, a discriminator model, and interactions between the generator and discriminator.

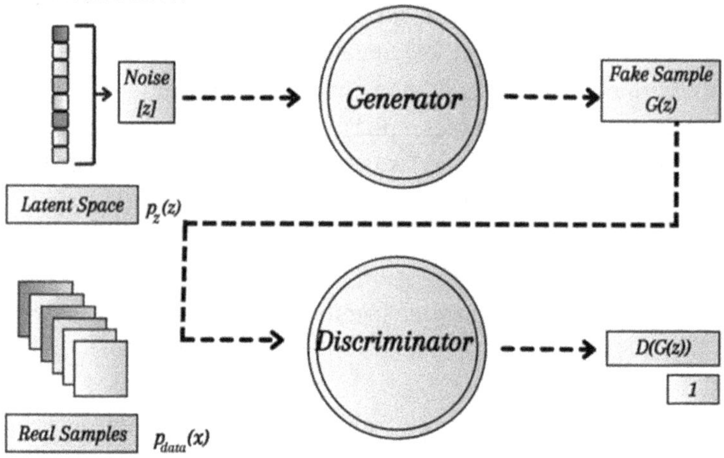

2.2 Advanced GANs

2.2.1 Deep Convolutional GAN (DCGAN)

Deep Convolutional GANs (DCGANs) solve constraints of vanilla GANs, especially in image production. They harness the power of Convolutional Neural Networks (CNNs) in both the generator and discriminator. CNNs excel in understanding the spatial relationships between pixels in images, vital for realistic image production(Radford et al., 2015). In the discriminator, CNNs learn features and exploit these to discern between actual and fraudulent images. Spatial information is modified by CNNs in the generator, effectively turning random noise into a real image layer by layer. Some specialized design decisions go into DCGANs in optimizing their performance. These include no use of pooling layers and fully connected layers. As a result, it generates high-quality and realistic images with more stable training compared to vanilla GANs. This breakthrough thus opened up the floor for additional improvements of GAN designs in many genres as shown in **fig 4**.

Figure 4. DCGAN Architecture for Image Generation: The design of a Deep Convolutional GAN emphasizes its convolutional layers, therefore being extremely ideal for producing realistic images.

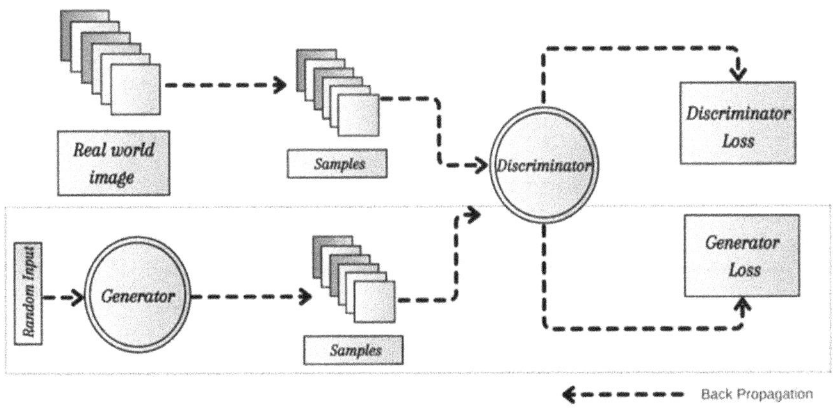

2.2.2 Conditional GAN (cGAN)

(cGANs) take the notion of GANs a step further by introducing additional control over the generated data. Unlike vanilla GANs or DCGANs that rely just on random noise, cGANs contain an additional input a condition. This condition can be anything from a category name (e.g., “cat,” “dog”) to a textual description (e.g., “a portrait of a smiling woman with red hair”).

This extra input permits the cGAN to provide data tailored to the precise situation presented. Imagine you want the generator to make images of different dog breeds. In a vanilla GAN, you'd get random dog-like outputs. But with a cGAN, you can input in the label “Golden Retriever” along with the noise, and the generator will use that information to build an image that closely resembles a Golden Retriever, with its characteristic traits and fur color.

This ability to regulate the generated data using labels or descriptions makes cGANs excellent for diverse applications. For example, cGANs can be used to generate realistic representations of new fashion items based on a certain style description, or build variations of existing products according on user preferences. They can also be used to provide realistic-looking medical data to train healthcare AI models. The possibilities are many, and cGANs continue to be a strong tool for creating data with a high degree of control and customisation as shown in **fig 5**.

Figure 5. Workflow of a Conditional GAN (cGAN: A depiction illustrating how cGANs produce pictures conditioned on extra input data such as class labels or characteristics.

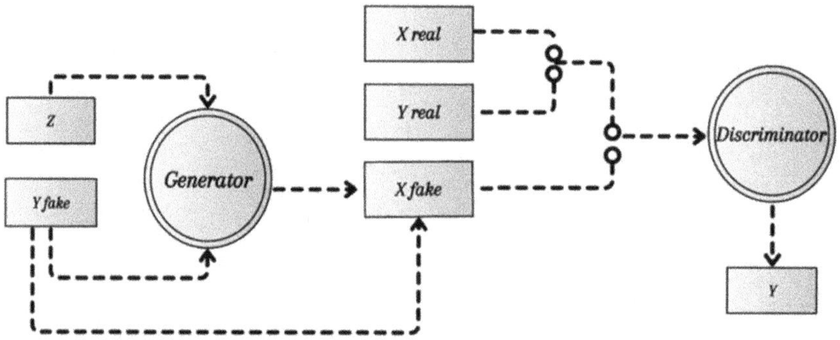

2.3 Specialized GANs

2.3.1 Super-Resolution GAN (SRGAN)

Super-Resolution Generative Adversarial Networks (SRGANs) specialize in one essential task: increasing image resolution. Imagine possessing a hazy, pixelated snapshot from your grandparents' age. An SRGAN can take that low-resolution image as input and generate a high-resolution rendition, bringing features and crispness back to life. This makes SRGANs important for applications like image restoration and video upscaling(Ledig et al., 2016).

Here's how SRGANs do this feat: Just with other GANs, SRGANs consist of a generator and a discriminator. The generator functions like a detail magician. It gets a low-resolution image and analyzes it to identify the underlying forms, textures, and missing information. Then, it employs its knowledge to continually build upon the image, adding details and boosting the resolution step-by-step.

The discriminator plays a crucial role in ensuring the generated image is not just upscaled but also realistic. It compares the high-resolution output from the generator with real high-resolution images. If the discriminator can't distinguish the generated image from a real one, it means the SRGAN has successfully enhanced the resolution while preserving image fidelity.

This adversarial training process between the generator and discriminator refines the SRGAN's ability to create high-resolution versions that are not only sharper but also believable. This makes SRGANs a powerful tool for restoring old, cherished photos, breathing new life into them, and for upscaling videos, allowing viewers to enjoy them in higher quality as shown in **fig 6**.

Figure 6. SRGAN Architecture for High-Resolution Image Generation: Overview of the Architecture of a Super-Resolution GAN, which creates high-resolution pictures from low-resolution inputs.

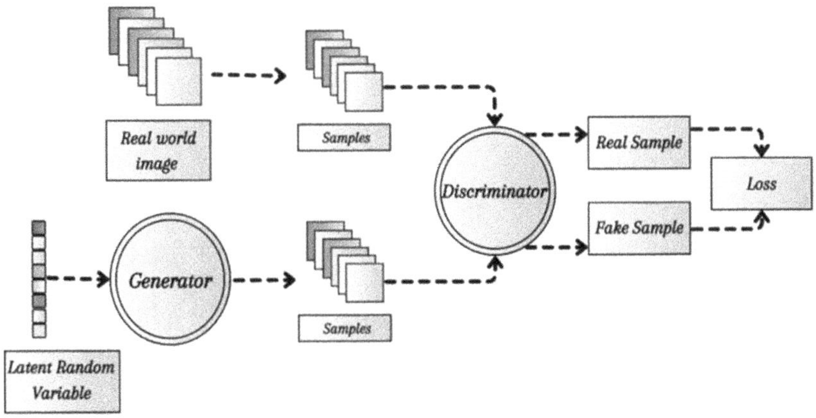

2.3.2 CycleGAN

CycleGAN is also called Pix2Pix GANs, are masters of image translation. Unlike standard GANs that focus on generating entirely new data, Pix2Pix GANs excel at transforming images from one domain (style, content) to another. Imagine having a photo of a majestic horse. A Pix2Pix GAN can take that image and, with the help of paired training data, convert it into a photorealistic image of a zebra, preserving the horse's pose and structure but applying zebra-like stripes. Similarly, it can transform a image of a sunny summer landscape into a snowy winter scene, complete with a realistic layer of white covering the ground and trees(Zhu et al., 2017).

This is a paired-image training dataset magic trick. Pix2Pix GANs are trained on datasets where every image has a corresponding other image in the target domain. For example, these may be photographs of horses and zebras in the same position. By evaluating these pairs, the Pix2Pix GAN learns the mapping between the two domains. It employs this knowledge to transform an input image from one domain for instance, horse to the matching image of the target domain for example, zebra.

The discriminator, as always, plays a key role. It ensures the generated image in the target domain (zebra) not only represents the content of the input image (horse) but also seems realistic and convincing within the target domain (realistic zebra stripes). This continual back-and-forth between the generator and discriminator refines Pix2Pix GAN's capacity to translate images precisely and realistically, making it a powerful tool for numerous creative applications as shown in **fig 7**.

Figure 7. Cycle GAN design for image-to-image translation: A CycleGAN is a unique architecture to learn mappings between two picture domains without paired data, enabling applications such as style transfer.

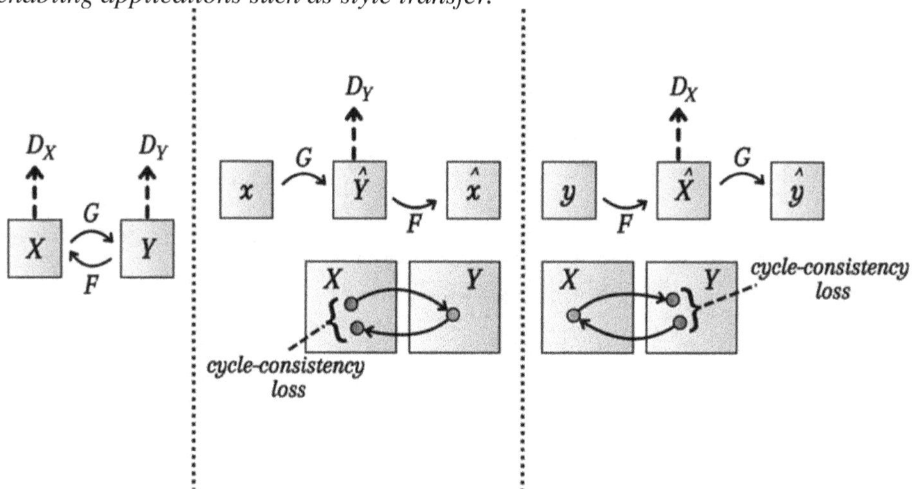

2.3.3 StyleGAN

StyleGAN2 stands out as a powerhouse in the realm of GANs, famous for its capacity to generate very realistic and high-fidelity images. It exceeds its predecessors by overcoming certain important weaknesses. Unlike prior GANs that occasionally battled with artifacts or unclear outputs, StyleGAN2 contains significant enhancements(Karras et al., 2018).

One of the important advances is that weight demodulation replaces adaptive instance normalization. Not having normalization layers allows better control to the generator regarding how an image is created, leading to sharper and more detailed outputs. This brings up another significant difference: StyleGAN2 does without the gradual, developing approach of prior models. While progressive growth starts with low-quality images and gradually improves the resolution, it generates irregularities. StyleGAN2 fixes this by adding a new training system, achieving high resolution without these difficulties.

Moreover, It contains certain unique regularization approaches in StyleGAN2 such as lazy regulation and path length regulation. These strategies allow the model to learn in a more regulated and stable fashion, minimizing difficulties such as unrealistic details or mode collapse (the generator is stuck creating few images which are often extremely identical).

The result of these developments is quite astonishing. StyleGAN2 can generate realistic images with rich details, particularly in areas like facial characteristics. It can catch tiny variations in style, enabling for the construction of highly diversified and realistic outputs. This makes StyleGAN2 a powerful tool for different applications, from generating high-quality product photos to making realistic portraits for artistic projects.

2.3.4 Text-to-Image GANs

Text-to-Image GANs: These networks, or Text2ImageGANs, transform your words into a graphic form. While normal GANs generate random images, Text2Image GAN will generate text based on an image concept. Like describing a sight as beautiful, Spectacular, and majestic waterfall tumbling from impossibly green looking mountains(Liao et al., n.d.; Ruan et al., 2021; Tan et al., 2023). A Text2Image GAN takes this text description as input and employs its deep learning prowess to generate a comparable image. Here's how it works:

At the heart of a Text2Image GAN are two models: a strong generator and an adversarial discriminator. A generator operates just as a visual storyteller. It initially parses the textual description, isolating significant pieces and ideas. The tool then uses this knowledge to build a visual rendering in sequence, one image "slice" at a time(Reed et al., 2016).

It is the obligation of the discriminator at all times to guarantee that the generated image corresponds to the specified text description. It therefore evaluates a created image against the authentic photos along with their descriptions, comparing all of them for probable differences. This will, therefore, constantly pushing the generator to develop its capacity to translate text into realistic, coherent visuals through this adversarial training cycle. After several training rounds, it becomes very good at reading text descriptions and generating related image s. Therefore, Text2Image GANs can be applied to applications like generating product image from descriptions or developing concept art based on textual narratives as shown in **fig 8**.

Figure 8. GAN architecture for description-to-image synthesis overview of the text-to-image GAN process, which creates pictures based on textual descriptions and displays the translation from semantic information to visual representation.

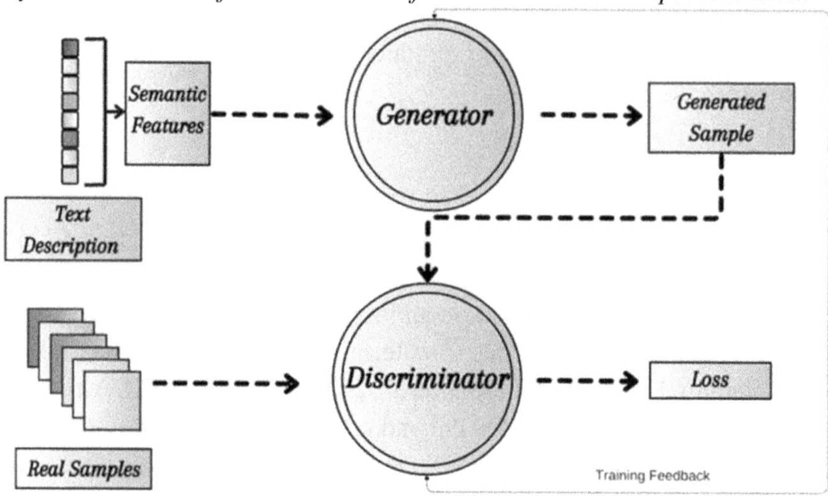

This **table 2** provides a comprehensive overview of different Generative Adversarial Network (GAN) architectures, highlighting their key strengths and potential applications. GANs are a powerful class of deep learning models that can generate realistic and creative data, from images and videos to text and music. By understanding the various types of GANs and their capabilities, we can explore their potential to revolutionize various fields.

Table 2. A survey of generative adversarial networks (GANs): Types, strengths, and applications

Category	Type of GAN	Description	Key Strength	Example Application
Foundational	Vanilla GAN	Original architecture with generator creating data from noise and discriminator evaluating its realism.	Simple to understand, foundational concept	Getting started with GANs
Advanced	Deep Convolutional GAN (DCGAN)	Utilizes CNNs in both generator and discriminator for superior image processing.	Generating high-quality, realistic images	Creating photorealistic images
Advanced	Conditional GAN (cGAN)	Allows control over generated data through additional input like labels or captions.	Tailored image generation	Creating images based on specific categories or descriptions
Specialized	Super-Resolution GAN (SRGAN)	Enhances image resolution by taking a low-resolution image and generating a high-resolution version.	Image upscaling and restoration	Restoring old photographs or upscaling videos
Specialized	CycleGAN	Translates images from one domain to another (e.g., horse to zebra, summer to winter).	Image-to-image translation	Creative image editing and style transfer
Specialized	StyleGAN	Known for exceptional realism and high-fidelity image generation, capable of capturing detailed features and style variations.	Unmatched image realism and detail	Creating highly realistic portraits or artistic imagery
Specialized	Text-to-Image GAN (Text2Image GAN)	Bridges the gap between words and visuals by creating images based on textual descriptions.	Generating product illustrations from descriptions, creating concept art from textual narratives	Text-to-Image GAN (Text2Image GAN)

3. POPULAR PUBLIC DATASETS

3.1 ImageNet: The King of Image Classification

Description: ImageNet is a titan in the realm of image datasets. It contains millions of photos precisely organized into thousands of object classifications. Imagine a digital collection featuring photographs of everything from airplanes and antelopes to zebras and zippers(*ImageNet*, n.d.).

Why it is helpful for GANs: ImageNet gives a great amount and variation of data; this makes ImageNet an invaluable resource specifically for Training Generative Adversarial Networks (GANs). The generator induces these features by training on millions of tagged samples for different things such as faces, automobiles boats etc. The consequence is that it can make very photoreal, state-of-the-art photographs covering hundreds of categories.

Applications: Trained on ImageNet, GANs can generate photorealistic product photos for e-commerce, create unique textures for 3D modeling, or even establish totally new artistic styles inspired by the actual world.

3.2 CelebA: Unveiling the Secrets of Faces

Description: Unlike ImageNet's concentration on objects, CelebA hones in on human faces. It's a big dataset specifically featuring celebrity faces with numerous features precisely annotated. These features include details like spectacles, stance, and even whether the person is smiling(*CelebA Dataset*, n.d.).

Why it's ideal for GANs: CelebA is a goldmine for training GANs to generate realistic and diversified portraits. The labeled properties help the GAN to recognize the intricacies of facial features, expressions, and accessories. This permits it to construct portraits with certain qualities, including producing photos of persons with eyeglasses or varied hairstyles.

Applications: GANs trained on CelebA can be used for creating realistic profile pictures, generating characters for video games with specific facial features, or even helping develop tools for facial recognition technology.

3.3 CIFAR-10 and CIFAR-100: Stepping Stones for Beginners

Description: CIFAR-10 and CIFAR-100 provide a friendlier introduction to image datasets compared to giants such as ImageNet. These datasets comprise smaller, lower-resolution photos (32×32 pixels), yet nonetheless encompass a range of object classifications. CIFAR-10 has 10 categories (e.g. airplanes and cars), whereas CIFAR-100 offers 100 categories(*CIFAR-10 and CIFAR-100 Datasets*, n.d.).

Why they are ideal for GANs (for beginners): Due to their modest size, CIFAR-10 and CIFAR-100 are suitable beginning places for people new to training GANs. They allow you to experiment with GAN designs and training strategies without the huge processing resources required for larger dataset.

Applications: Although not optimal for generating extremely realistic images owing to the lower resolution, these datasets can be useful for studying GANs' ability of GANs to learn and generate simpler objects. They can act as stepping stones before moving into larger and more sophisticated datasets.

3.4 Flickr and Places365: Painting the World with GANs

Description: Flickr and Places365 take you beyond items and faces, entering the world of scenes. These databases include enormous collections of photos covering various locales and habitats. Imagine a digital travelogue with photographs of bustling cityscapes, tranquil beaches, or spectacular mountains(*Places365 Dataset | Papers With Code*, n.d.).

Why they're perfect for GANs: Flickr and Places365 provide the perfect training ground for GANs specializing in generating realistic landscapes or outdoor scenes. By analyzing these images, the generator learns the relationships between sky, land, water, and vegetation, allowing it to create believable and diverse outdoor environments.

Applications: GANs trained on these scene datasets can be used for producing virtual backgrounds for video conferencing, making realistic backdrops for animation or movies, or even designing applications for urban planning and landscape design.

This **table 3** explores the synergy between various image datasets and Generative Adversarial Networks (GANs). Selecting the right dataset is crucial for effective GAN training, as it provides the raw material from which the network learns to generate novel and realistic data. Each dataset offers unique advantages for specific GAN applications.

Table 3. Image datasets: Fueling the creativity of generative adversarial networks (GANs)

Dataset	Description	Why Perfect for GANs	Applications
ImageNet	Massive collection (millions of images) categorized into thousands of object classes	Unmatched size and diversity provide vast information for GANs to learn, enabling generation of highly realistic and varied images	Photorealistic product images, unique textures for 3D modeling, new artistic styles
CelebA	Large dataset of celebrity faces with labeled attributes (glasses, pose, smile)	Enables GANs to understand facial features, expressions, and accessories, allowing generation of portraits with specific characteristics	Realistic profile pictures, video game characters with specific features, facial recognition technology development
CIFAR-10 & CIFAR-100	Smaller datasets with lower-resolution images (32x32 pixels) containing various object classes (10 or 100 categories)	Manageable size for beginners to experiment with GAN architectures and training techniques	Explore GANs' ability to learn and generate simpler objects, stepping stone for larger datasets
Flickr & Places365	Vast collections of scene images depicting diverse locations and environments	Ideal training ground for GANs specializing in landscapes or outdoor scenes, allowing creation of believable and diverse environments	Virtual backgrounds for video conferencing, realistic backdrops for animation/movies, urban planning and landscape design applications

Choosing the Right Dataset:

1. Task Specificity: Consider the precise type of photographs you wish to make. Choose a dataset that fits with your desired outcome (e.g., portraits, landscapes, specific objects).
2. High-Quality Data Set: Provide a dataset of high-quality images with a minimum quantity of noise/artifacts. This aids the learning of correct representations in a GAN and demonstrates realistic visuals.
3. Data amount: The amount of the dataset might effect training time and results. Larger datasets often lead to better outcomes but require more processing resources. Balance these parameters based on your project needs and capabilities.
4. Data Labeling: Consider datasets that are appropriately labeled or annotated for tasks in which one needs control over the generated images, such as with cGANs, and this supplementary information would be provided as input during training.

4. APPLICATION OF GAN

Generative Adversarial Networks (GANs) have transformed the field of image production. Their ability to create high-fidelity and realistic images has opened up a huge array of applications, impacting numerous creative and practical fields. Here's a closer look at how GANs are being used in image generation:

4.1 Creative Applications

4.1.1 Art Generation

GANs are consequently offering artists a means to create entirely new and unique pieces, generating a storm in the art world. They can be considered artistic springboards in the following ways: First, GANs might come to the heart of artistic styles. Artists who feed them data sets of paintings from a certain artist or movement may develop new works that echo those styles. This gives freedom for experimentation in which paintings can be created in the styles of artistic idols or even a cross between styles from different movements. On the other side, GAN functions as a way of conceptual exploration: an artist feeds on the idea of "dreamscapes," and GAN will surprise one with images expressing that precise idea. To this purpose, it may establish new creative trends and artistic pushing of boundaries. Mods employing GANs allow for variations of prior works. One may therefore feed an artifact of an artist to a GAN and get back additional replicas with varied colors, compositions, or details. It is in this sense that it opens up the chance for an artist to examine things anew, perhaps setting out some unbaked creativity within. It is worth emphasizing that GANs are not a replacement for an artist but a supplement to the arsenal of their creativity. The artist stands at the center of the whole process: picking and curating the training data so that the GAN will learn from the correct styles, enhancing that output with an artistic vision of his own, and offering a conceptual framework or storytelling element where GANs, brilliant alone, fall somewhat flat. But in their visual effects, they could not manage on their own. In other words, GANs have grown as a medium of extraordinary power that allows an individual to transcend the specified bounds given by art and develop new paths of creativity. They can merge data and artwork in the hands of the right artist.

4.1.2 Product Design

The new role that GANs are playing in product design is similar to being a digital design powerhouse. Designers can use them to create highly realistic photos of a product prototype instantly, enabling acceleration of the design cycle. They can further be used to modify existing designs to look at other textures, finishes, and colors. GANs can also be used to compose marketing mockups demonstrating products in use scenarios all within the digital environment to enhance creativity, drive efficiency, and launch compelling products.

4.1.3 Fashion Design

GANs are entering the industry connected to fashion creation by delivering a digital fabric cutter and tailor in one. They can be utilized for making new garment patterns by creating fresh motifs and textures. Older designs can receive a second life by giving a GAN a base garment and letting it come up with variations that differ by collar, sleeve, or ornamentation. GANs may, however, generate realistic photographs of models in such novel designs; these are variegated in fabrics and colors. This consequently, allows designers to fiddle electronically with such a big number of fashion concepts before they have to touch a textile physically, thereby shortening the design process and generating an all-new sense for creative exploration in this business.

4.2 Practical Applications

4.2.1 Image Editing:

GANs are essentially a digital artist's tool in image alteration. They can execute magnificent deeds for example, inpainting: repairing areas of an image that are lacking realistic qualities. Imagine restoring that ripped family photo with GANs quickly filling in the missing part. Even old or damaged photographs are given a new lease of life via photo restoration utilizing GANs. After incrementing quality and removing faults, the faded memories are brought back into magnificent detail(Niu et al., 2020).

4.2.2 Medical Imaging

Medical imaging is awash in the current of GANs because they produce synthetic medical images. Notably, this works best with really scarce medical data or privacy-related issues. These synthetic images could be used for training AI algorithms in medical diagnosis and treatment planning to yield perfect quality training datasets without any violation of patient privacy or requiring enormous volumes of accurate data. That may, in turn, lead to more sophisticated and accurate AI tools for medical professions(Singh & Raza, 2021).

4.2.3 Media and Entertainment

Apart from the artistic arena, GANs are working their magic within cinema and video games. They can also generate really impressive special effects that make the scenario come to life. Imagine a vast motion picture in which GANs generated a realistic dragon that flies through the sky. In addition, GANs can be used in the construction of great backgrounds or even the generation of virtual life by populating virtual worlds with lifelike inhabitants. Further creating immersive and visually stunning experiences, pushing the frontiers of what would be possible in movies and video games.

This **table 4** showcases the remarkable versatility of Generative Adversarial Networks (GANs) across various creative and practical domains. By harnessing the power of GANs to generate new and realistic data,

Table 4. The power of GANs: Unleashing creativity and practical applications

Category	Application	Description	Example
Creative	Art Generation	Create entirely new and unique artwork	Generate paintings in the style of Van Gogh or create abstract art pieces.
Creative	Product Design	Visualize product prototypes and design variations	Generate mockups of new phone designs or explore different color schemes for furniture.
Creative	Fashion Design	Design new clothing patterns or create photorealistic images of models	Generate variations on existing dress designs or create images showcasing different outfit combinations.
Practical	Image Editing	Inpaint missing parts of images or restore old photographs	Enhance blurry photos or fill in cracks in historical paintings.
Practical	Medical Imaging	Generate synthetic medical images for training purposes	Create realistic MRI scans for training AI algorithms in disease detection.
Practical	Media and Entertainment	Create special effects in movies and video games	Generate realistic explosions or populate virtual worlds with diverse characters.

Beyond these, here are some emerging applications:

1. Satellite Imagery: The application of GANs is used to raise the amount of detail in low-resolution satellite photos. These AI models work like detectives, trained on fuzzy and clear image pairings to fill in the missing information, thus providing clearer, high-resolution photos that will be useful for advanced environmental monitoring and better catastrophe response.

2. Augmented Reality: GANs are sculptors for AR objects. They are trained on piles of real-world item data and create hyper-realistic 3D models in virtual space, thereby infusing life into AR experiences where believable furniture, monsters, or anything imagined come right before your very eyes.

The potential of GANs in image generation is immense. These are obviously going large, and we should expect many more creative uses to emerge shortly. It's vital to emphasize that there is also a duty to the progress of GAN technology with reference to ethical considerations about prospective exploitation for things like deepfakes.

5. IMPACT OF GAN IN IMAGE GENERATION

Indeed, the effect that GANs have had on the art of image creation can only be regarded as really revolutionary. This outlines some of the major areas where GANs are showing to have quite an impact:

1. Democratization of Image Creation: GANs make it conceivable for any user to create a high-quality image independent of their artistic prowess. This allows creative expression at the individual level and corporations' ability to develop visuals with minimum design resources conceivable.
2. Higher creativity: With GANs, artists might create a new type of art, variants on current concepts, and could even team up with the AI to come up with complete new pieces.
3. Product Development: GANs can give greater vigor to product design by prototyping visualization and checking out variations in design; it will also generate marketing material with only realistic mock-ups of products.
4. different Advances: GAN works on artificial medical images that can be helpful for different other medical studies and facilitate diagnosis. In addition, GAN could produce photo-realistic visual effects for movies and games or make high-resolution satellite photos for environmental monitoring.

Challenges and Considerations:

1. Deepfakes and disinformation: The simplicity with which near-photorealistic and super-real fake images may be made raises concerns concerning the development of deepfakes for disinformation or destroying reputations.

2. Ethical concerns: The possible misuse of GANs calls for rigorous consideration regarding ethical issues and responsible developmental methods.
3. Bias in training data: GANs can perpetuate any bias in the training data. Efforts should be placed to remove prejudice to guarantee fair and responsible use of these models.

Overall Impact:

Despite being counter to common logic, GANs are the invention that will etherealize image generation. The high-fidelity graphics generated will spark imagination, boost several areas of a person, and empower the individual. However, with it comes ethical challenges to be engaged in and addressed with to guarantee responsible improvement and use of this powerful innovation. As research into GANs continues to advance, we may predict much more breakthrough applications to rise, impacting the future of picture creation past.

6. CONCLUSION

This chapter has addressed the interesting field of generative adversarial networks and the change they have in store for image generation. We start our trip by showcasing how creative visual content is materialistic in many fields, all with expanding demand. Traditional technologies in image generation, albeit practical, were bound by realism and the flexibility to create. GANs came as a major game-changer with their adversarial training technique.

We discussed the essential concepts of GANs about conventional techniques. The pioneering GANs set the foundation from basic designs, while advanced designs the DCGANs, cGANs, and StyleGANs moved the frontiers related to realism and control over images. Specialized GANs, like the SRGAN for picture upscaling and Text-to-picture GAN, further underlined the adaptability of this approach. Publicly popular datasets such as ImageNet and CelebA were explored as key materials for training GANs.

The final section of the chapter focused on the numerous applications of GANs. From the creative domain of art generation and fashion design to the practical world of medical imaging and media special effects, GANs are influencing numerous fields. We investigated how GANs are not just introducing new opportunities but also streamlining workflows and boosting existing procedures.

The impact of GANs on image generation is apparent. They have opened doors to previously inconceivable levels of realism and creative investigation. However, obstacles and issues remain. As GANs get more complex, ethical problems regarding deepfakes and associated biases deserve serious attention. Responsible development and utilization will be essential to ensure GANs are a force for good.

Looking forward, the future of GANs seems promising, particularly in a few important areas of study. Advancing the state of the art in enhancing control over produced material-for example, being able to change particular properties in generated images at fine-grained detail-will open up new application possibilities in media, design, and customized content production. Work on decreasing the computational complexity of GANs may make them more accessible and, consequently, capable of broader deployment in resource-constrained situations such as mobile devices or edge computing.

Meeting these hurdles and grabbing the potential, GANs may totally revolutionize the way we create, visualize, and even interact with the environment, assuring future innovation in varied businesses and research disciplines.

REFERENCES

"CelebA Dataset." Accessed: May 28, 2024. [Online]. Available: https://mmlab.ie.cuhk.edu.hk/projects/CelebA.html

"CIFAR-10 and CIFAR-100 datasets." Accessed: May 28, 2024. [Online]. Available: https://www.cs.toronto.edu/~kriz/cifar.html

Goodfellow, I., Pouget-Abadie, J., Mirza, M., Xu, B., Warde-Farley, D., Ozair, S., & Bengio, Y. (2020). Generative adversarial networks. *Communications of the ACM*, *63*(11), 139–144.

Gu, J., Shen, Y., & Zhou, B. (2020). *Image Processing Using Multi-Code GAN Prior*. DOI: 10.1109/CVPR42600.2020.00308

"ImageNet." Accessed: May 28, 2024. [Online]. Available: https://www.image-net.org/

Karras, T., Laine, S., & Aila, T. (2018, December). A Style-Based Generator Architecture for Generative Adversarial Networks. *IEEE Transactions on Pattern Analysis and Machine Intelligence*, *43*(12), 4217–4228. DOI: 10.1109/TPAMI.2020.2970919 PMID: 32012000

Ledig, C., Theis, L., Huszár, F., Caballero, J., Cunningham, A., Acosta, A., & Shi, W. (2017). Photo-realistic single image super-resolution using a generative adversarial network. In *Proceedings of the IEEE conference on computer vision and pattern recognition* (pp. 4681-4690).

Liao, W., Hu, K., Yang, M. Y., & Rosenhahn, B. (2022). Text to image generation with semantic-spatial aware gan. In *Proceedings of the IEEE/CVF conference on computer vision and pattern recognition* (pp. 18187-18196).

Mirza, M., & Osindero, S. (2014). Conditional generative adversarial nets. arXiv preprint arXiv:1411.1784.

Niu, S., Li, B., Wang, X., & Lin, H. (2020, July). Defect Image Sample Generation with GAN for Improving Defect Recognition. *IEEE Transactions on Automation Science and Engineering*, *17*(3), 1611–1622. DOI: 10.1109/TASE.2020.2967415

"Places365 Dataset | Papers With Code." Accessed: May 28, 2024. [Online]. Available: https://paperswithcode.com/dataset/places365

Radford, A., Metz, L., & Chintala, S. (2015). Unsupervised representation learning with deep convolutional generative adversarial networks. arXiv preprint arXiv:1511.06434.

Reed, S., Akata, Z., Yan, X., Logeswaran, L., Schiele, B., & Lee, H. (2016, June). Generative adversarial text to image synthesis. In *International conference on machine learning* (pp. 1060-1069). PMLR.

Ruan, S., Zhang, Y., Zhang, K., Fan, Y., Tang, F., Liu, Q., & Chen, E. (2021). Dae-gan: Dynamic aspect-aware gan for text-to-image synthesis. In Proceedings of the IEEE/CVF international conference on computer vision (pp. 13960-13969).

Singh, N. K., & Raza, K. (2021). Medical Image Generation Using Generative Adversarial Networks: A Review. *Studies in Computational Intelligence*, *932*, 77–96. DOI: 10.1007/978-981-15-9735-0_5

Tan, H., Liu, X., Yin, B., & Li, X. (2023, December). DR-GAN: Distribution Regularization for Text-to-Image Generation. *IEEE Transactions on Neural Networks and Learning Systems*, *34*(12), 10309–10323. DOI: 10.1109/TNNLS.2022.3165573 PMID: 35442894

Zhu, J. Y., Park, T., Isola, P., & Efros, A. A. (2017). Unpaired image-to-image translation using cycle-consistent adversarial networks. In *Proceedings of the IEEE international conference on computer vision* (pp. 2223-2232).

Chapter 7
Fundamentals of GANs

Manjushree Nayak
https://orcid.org/0000-0001-6383-780X
NIST University, India

Subhakanta Panda
NIST University, India

ABSTRACT

In 2014, Ian Goodfellow and colleagues introduced Generative Adversarial Networks (GANs), transforming AI by enabling the creation of highly realistic synthetic data, including text, music, and images. GANs consist of two neural networks: the generator, which creates data samples starting from random noise and improves them iteratively, and the discriminator, which distinguishes between real and generated data, providing feedback to enhance the generator's realism. This adversarial process drives both networks to improve until the generator produces high-quality synthetic data. GANs excel in representing complex data distributions without explicit probabilistic models, ideal for tasks like image generation. However, GAN training is challenging, with issues like training instability and mode collapse. Researchers have developed strategies like regularization techniques and architectural adjustments to address these problems. Despite challenges, GANs are widely used in computer vision, natural language processing, music creation, drug discovery, and privacy preservation.

INTRODUCTION

Generative Adversarial Networks (GANs), launched by Ian Goodfellow and his team in 2014, have become one of the most innovative and influential machine learning frameworks in the field of artificial intelligence. The generator's role is to

DOI: 10.4018/979-8-3693-7152-7.ch007

create artificial data, such as images, videos, or text, that resembles real-world data. While the job of the discriminator is to differentiate between real data and generated data. This setting promotes a competitive environment. where both networks update iteratively in generations The basic idea is rooted in game theory. Where the generator aims to minimize the loss by deceiving the discriminator. And the discriminator will try to maximize accuracy.

Figure 1. Fundamentals of GANs

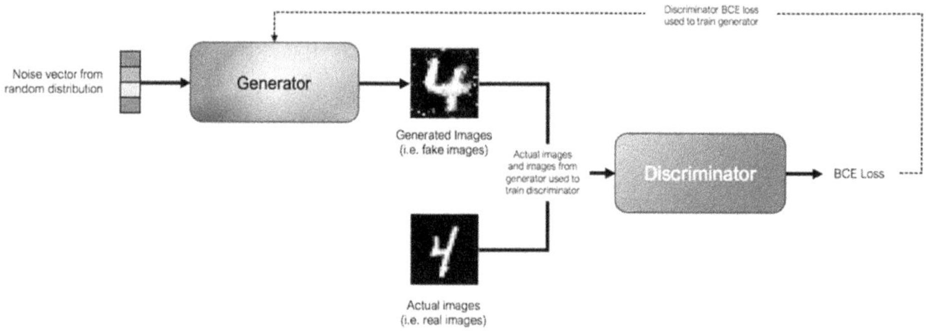

GANs have shown transformative potential in many fields. in computer vision It is widely used in image synthesis. The resolution is too high. and format settings This makes it possible to use various applications, such as creating realistic images from text descriptions. or improving low-resolution images in health care Generating high quality synthetic data to improve GAN training, non-sampling datasets or MRI scans for better diagnosis. They have helped in reconstructing medical images. GANs in entertainment and creative arts have revolutionized the creation of realistic characters. Video game environment original artwork and song composition and application of data updates Addressing the challenges of limited data sets in areas such as autonomous driving and natural language processing. (M. S., et al. 2022).

Despite their potential, GANs have significant challenges. One of the most important problems is training instability. where the constructor and the discriminator struggle to achieve a stable balance. This leads to suboptimal results. Another problem is the collapse of the mode. The generator creates a limited variety of data. and it is not possible to capture the full diversity of the basic data set. Researchers have proposed various solutions. So much for these problems. This includes advanced loss functions such as the Wasserstein loss, architectural innovations such as Deep Convolutional GANs (DCGANs), and advanced training techniques such as feature matching and gradient banding.

In addition to technical boundaries, Gans also increases moral concerns. In addition to creating hyper-realistic depths, it can be used to manipulate media and spread wrong information. Exposing the need for the outline and detection mechanisms of strong governance, trained guns on biased datasets inadvertently eliminate social stereotypes. This requires efforts to diversify and cure training data responsibly. The future is full of promise. Researchers are discovering integration with other state-of-the-art technologies such as autopylals to improve text-image synthesis. For quantum computing fast training, in addition, the development of centralized GANS for privacy protection and adaptive GANs that can learn in real time. Development presents exciting opportunities because GANS continues to develop. And they are ready to join individual healthcare to scientific research and further industries, while stimulating the need for moral inspection to ensure proper use. Responsible GANS represents groundbreaking advance in generative modeling. It provides unique abilities in generating synthetic data. Their influence is in many areas, from art and entertainment to healthcare and cyber security. This makes them a mark in the continuous development of artificial intelligence generators network with the goal of generating the same data as real data. Data is the end result of starting random noise through several levels of sample change. In contrast, the classification network acts as a classifier that separates the data samples generated by the generator from real samples. Both networks will get better with frequent training: discrimination will become better in detecting fake data. And the generator will be better in the generate. (M. S., et al. 2022).

APPLICATIONS AND IMPORTANCE

Gans have an impact on a wide range of regions and applications, proving their adaptability and importance in contemporary technology. Major uses for Gans are:

Image generations: Gans have applications in entertainment, media and arts because they can create incredibly realistic images from scratches. For example, the beautiful landscape, artwork, and even the life representative of non-existent individuals is produced using Gans.

Data hike: GANS can produce synthetic data to complement the training dataset when obtaining sufficient versions of labeled data. This machine helps in increasing the performance of the learning model, especially in domains such as medical imaging where data is often rare (Nayak, M., et al. 2023).

Super-resolution: GAN is used to improve image resolution by converting low-resolution input into high-resolution output. This is a significant implication for improvement in safety, satellite photography and medical diagnosis image quality.

Style transfer: Gans enable creative applications such as convenience of the transfer of artistic style from one image to another, turning photos in pictures in the similarity of famous artists. Video generations and animation: Gans are useful in film and video game industries as they can produce video frames, conscious characters and produce realistic motion sequences.

Healthcare: By producing molecular structures, GANS can help in the development of drugs and even produce synthetic medicine for research purposes. Text-to-Image synthesis: Gans are capable of producing images from written details. It has applications in the areas of design, marketing and e-commerce as it allows for the creation of visual material from text input. (Fan, P., et al. 2022).

Medical Imaging

Case Study: Enhancing MRI Scans with GANs: GANs have been instrumental in improving the resolution and quality of medical images. For example, GAN-based models such as pix2pix and CycleGAN have been used to enhance low-resolution MRI or CT scans, enabling clearer visuals for accurate diagnosis. A notable application is in reconstructing high-quality images from undersampled data, reducing scan time and patient discomfort. By preserving critical diagnostic details, these GAN-enhanced scans improve the accuracy of detecting diseases like cancer, neurological disorders, or cardiovascular issues.

Societal Impact: GANs contribute to better healthcare accessibility, especially in regions with limited access to advanced medical imaging equipment. By synthesizing realistic medical data for training AI models, GANs also enable researchers to develop diagnostic tools without the need for sensitive patient data, preserving privacy. (Reyntjens, S., et al. 2001).

Cybersecurity

Case Study: GANs for Malware Detection: In cybersecurity, GANs are used to simulate attack scenarios and improve defense mechanisms. For instance, researchers have leveraged GANs to create adversarial malware samples that test the robustness of security systems. By training defensive algorithms with these synthetic samples, companies can proactively identify vulnerabilities in their systems.

Societal Impact: GANs help build stronger cybersecurity infrastructures, protecting sensitive personal and corporate data from cyberattacks. They also aid in simulating and mitigating potential threats, contributing to a safer digital environment.

Creative Arts and Entertainment

Case Study: Generating Realistic Game Characters with StyleGAN: GANs have revolutionized the gaming and entertainment industries by enabling the creation of photorealistic characters and environments. NVIDIA's StyleGAN has been widely used to generate high-resolution human faces for video games and virtual environments. These faces are indistinguishable from real humans, adding a layer of realism to immersive experiences.
Societal Impact: GANs democratize creativity by providing tools for artists and game developers to generate unique assets without extensive resources. This enhances storytelling and visual experiences while reducing production costs and timelines.

Personalized Healthcare and Drug Discovery

Case Study: GANs for Synthesizing Drug Molecules: In pharmaceutical research, GANs like MolGAN are used to design novel drug molecules by generating new chemical structures. These GANs accelerate the discovery process by predicting molecular properties and optimizing candidates for drug efficacy.
Societal Impact: By reducing the time and cost of drug development, GANs have the potential to revolutionize medicine, making treatments more accessible and personalized to patients\u2019 needs. (Reyntjens, S., et al. 2001).

The importance of GANs comes from their capacity to provide high-quality data as well as from their contribution to the advancement of neural networks and deep learning research. GANs have created new opportunities for study and invention by expanding the realm of generative modeling.

The Core Concept

Gaining an understanding of GANs necessitates knowing the following basic ideas:

Adversarial Training: The fundamental aspect of GAN training is the interaction between the discriminator and generator. The networks are iteratively improved by this adversarial process, producing high-quality generated data.

Generator Network: This system converts random noise into data that is similar to the training data. It attempts to trick the discriminator in order to learn how to provide realistic outputs.

Discriminator Network: The discriminator differentiates between actual data from the training set and fictitious data generated by the generator, functioning as a binary classifier. Its input aids in the generator's development. (Shehadeh, M., et al. 2012).

Loss Functions: The generator and discriminator are trained based on their respective loss functions. The discriminator seeks to maximize the likelihood that one of its outputs will be accurately identified as phony, while the generator seeks to limit this likelihood.

Minimax Game: The generator and discriminator in a minimax game have competing goals. This is how GAN training works. When the discriminator is unable to distinguish between produced and actual data, the optimal state has been attained.

ARCHITECTURE OF GANs

Discriminatory and generators are two main neural networks that create architecture of generative adverse networks (GANS). These networks are involved in a dynamic adverse process in which the discriminatory wants to consider between the manufactured and real data, and the generator wants to generate realistic data. This block explains the complex design of the generator and the discrimination, how they work together during training, and the general dynamics of Gans. Generator network The generator network is in charge of producing fake data that repeats how to distribute real data. It creates an output that looks like a real data by changing an input that is a random noise vector through many layers. Usually, the architecture of the generator consists of the following elements: Layer of input: The generator begins with a vector of random noise, usually drawn from a normal or equal distribution. This noise vector is the basis for making data. Dense layers: The first layers of the generator are often fully connected, or densely, layers that help in making the input noise vector more dimensional. Non-linear activation functions, such as Relu (improvised linear unit), are used to introduce non-linearity after these layers. (Shehadeh, M., et al. 2012).

Deconvolutional Layers: Layers used to upsample the input to higher dimensions are called deconvolutional layers, also referred to as transposed convolutional layers. In order to get high-resolution outputs from low-dimensional noise vectors, deconvolutional layers are essential. Activation functions like ReLU or Leaky ReLU are frequently used by them.
Batch Normalization: After every deconvolutional layer, batch normalization is often employed to stabilize training and speed up convergence. It does this by taking the batch mean and dividing it by the batch standard deviation, which normalizes the output from the previous layer. (Nayak, M., et al. 2024).
Output Layer: The generator's last layer usually produces the output by applying a Tanh activation function to picture data. This output must fall between a predetermined range, for example, [-1, 1] for normalized image data.

Figure 2. Generator architecture

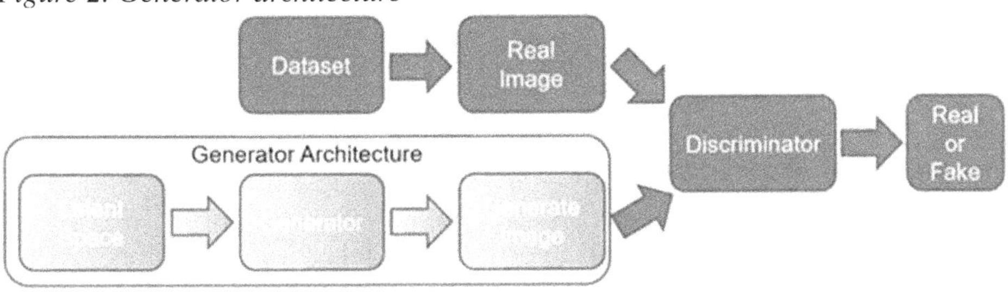

Discriminator Network

To differentiate between actual data samples and artificial data generated by generators, the discriminatory network serves as a binary classifier. Its architecture has specific components:

Input layer: Real or manufactured data is fed into the discriminatory as input.
Convisional layers: The first layer of the deciminator is the convenient layers, which take input data and extract hierarized features from it. To reduce the missing shield problem, these layers use activation functions such as leak relay to enable gradients to flow through non-active neurons.
Pooling layers: By reducing the spatial dimensions of data and capturing the most important properties, pooling layers - such as Max Pooling - Reduce Computational Efforts. **Batch normalization:** Used to stabilize training and accelerate convergence, batch normalization works similar to the generator.

Thick layers: Finally, fully connected layers combine recovered features to determine binary classification.

Output layer: The last layer often outputs a probability value between 0 and 1, indicating whether the input data is real (near 1) or fraud (close to 0). It uses a sigmoid activation function. (Cho, Y. K. R., et al. 2017).

Figure 3. Discriminator architecture

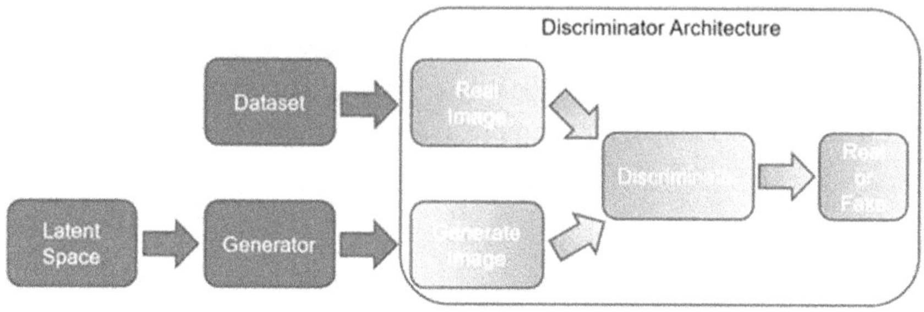

Dynamics of Training

In order to maximize each of their goals, the generator and discriminator in a GAN are trained by alternating updates. The following is a description of this process:

Train Discriminator:

Step 1: Take a sample of actual training set data.

Step 2: Use the generator to create fictitious data by sampling a set of noise vectors.

Step 3: Use both false and real data to calculate the discriminator loss.

Step 4: Use backpropagation to update the discriminator's weights in order to maximize the likelihood that real and fake data will be accurately classified.

Training Dynamics:

The generator and discriminator in a GAN are trained by alternating updates to optimize each of their objectives. This procedure is explained as follows:

Train Discriminator:

First, select some real training set data.

Step 2: Sample a set of noise vectors with the generator to produce fake data.

Step 3: Determine the discriminator loss by using both real and fake data.

Step 4: To increase the chance that actual and fake data will be correctly classified, use backpropagation to adjust the discriminator's weights. (Le-The, H., et al. 2021).

Training Challenges:

GAN training can be difficult for a number of reasons:

Convergence: Ensuring that the GAN reaches a stable state when high-quality data is produced by the generator and the discriminator finds it difficult to discern between actual and fake data.

Moderate Collapse: The generator may concentrate on a small number of modes of the data distribution and produce a restricted range of outputs. To solve this, a variety of strategies are used, including mini-batch discrimination and feature matching.

Vanishing Gradients: If the discriminator is overconfident, the gradients will disappear and the generator won't be able to learn efficiently. This issue can be lessened by using different loss functions, such as Wasserstein loss. (Lewis, S. M., et al. 2022).

The strong generative powers of GANs are based on their architecture, which combines discriminator and generator networks. Through an adversarial training process, these networks push each other to get better, producing artificial data that is remarkably lifelike. Gaining an understanding of GANs' intricate design and training dynamics is crucial to realizing their full potential in a range of applications and improving the field of generative modeling research.

MATHEMATICAL FOUNDATIONS OF GANS

An understanding of these networks' mathematical underpinnings is essential to understanding how and why Generative Adversarial Networks (GANs) work. The main mathematical ideas behind GANs are examined in this section, along with how loss functions are defined, how game theory functions, and typical problems like convergence.

Functions of Loss

The adversarial aspect of the discriminator and generator is captured by the loss functions in GANs. The concept of a two-player minimax game serves as the basis for these loss functions.

Loss of Discrimination

The discriminator's goal is to accurately discern between created (false) and actual data. The discriminator's loss function can be found by:

$$LD=-E\mathbf{x}\sim pdata[\log D(\mathbf{x})]-E\mathbf{z}\sim pz[\log(1-D(G(\mathbf{z})))]$$

Here:

$D\,(x)$

The discriminator's assessment of the likelihood that actual data xx is real is given by D(x).

$D\,(G\,(z))$

The discriminator's estimate of the likelihood that produced the data is D(G(z)).

The function G(z) (derived from the noise z) is real.

Ex∖ pd a b

E x p data

The predicted value across all genuine data samples is indicated by.

E z p z $E\ z \sim p\ z$

The predicted value across all noise samples, z, is indicated by.

By maximizing this loss function, the discriminator seeks to improve its capacity to discern between authentic and fraudulent input.

Loss of Generator

The generator's goal is to trick the discriminator into believing that the created data is authentic. The generator's loss function is provided by:

$J\,G = -E\ z \sim p\ z\ [\log\ \ D\,(G\,())\,]$ * L * G = –E z p z [logD(G(z))]

In this case, the generator aims to reduce the likelihood that the discriminator would accurately detect its output as fraudulent.

Minimax and Game Theory From the perspective of game theory, game GAN training can be understood as a minimax game between the discriminator and the generator. The formal goals of the discriminator and generator are as follows:

min G max $D\,V(D, G) = E\,x \sim p\,d\,t\,a$

The expression [log $D\,(x)$] + $E\,z \sim p\,z$ [log $(1 - D\,(\boldsymbol{D}\,(z)))$]

min I G I max I D I V(D,G) = E x I p data

[logD(x)] + E z \p z [log(1)−D(G(z))]

Here:

The generator G attempts to minimize the value function V(D,G) = $V(D, G)$.

The same value function is the goal of discriminator D.

This formulation accurately reflects the adversarial character of GANs, where the ideal solution is $D\,(G\,()) = 0.5$ D(G(z)) = 0.5 for every z z, where the discriminator is unable to discriminate between genuine and generated data.

Convergence Problems

A significant obstacle to convergence in GANs is the intricate relationship between the discriminator and generator. A number of important topics come up:

Mode Collapse: The generator concentrates on particular modes of the data distribution, yielding a restricted range of outputs. This may occur if the generator discovers a small number of modes that consistently trick the discriminator.

Disappearing Gradients: Very modest gradients for the generator can result from the discriminator classifying generated samples with high confidence if it gets too powerful. This makes it more difficult for the generator to grow and learn.

Non-Convergence: Rather than converge to a stable equilibrium, the training dynamics of GANs may cause oscillations or divergence.

Options and Remedies

Several changes to the conventional GAN framework have been suggested in order to overcome these problems:

Feature Matching: The generator is trained to match the statistics of the features taken from an intermediary layer of the discriminator, as opposed to using the discriminator's output directly. In order to keep mode collapse from happening, this encourages the generator to generate a variety of samples.

Minibatch Discrimination: This method helps identify a lack of diversity in the generated samples by adding additional layers to the discriminator so that several samples can be considered at once.

Differential Loss Functions: The Jensen-Shannon divergence is not used in Wasserstein GAN (WGAN); rather, the Earth-Mover (Wasserstein) distance is used. The WGANs' loss functions are as follows:

X **v** ~ *p d a t* **O**

[*D* (*x*)] − *E z* ~ *p z* [*D* (*G* (*z*))] L D = E x **p **data

[D(x)]−E z~p z [D(G(z))] *J G* = −*E z* ~ *p z* [*D* (*G* (*z*))] * L * G = −E z p z [D(G(z))]

Even with a well-trained discriminator, the Wasserstein distance yields more meaningful gradients, which promotes more steady training.

Gradient Penalty: In WGANs, a gradient penalty term is added to the discriminator's loss in order to enforce the Lipschitz constraint:

L **v** =*L D* +*λ E* **l** ^ ~ *p x* ^ [(∑ ∇ *x* ^ *D x* ^) || 2 − 1) 2 L D = L D + λE x^~p x^ [(||∇ x ^ ^ D(x^)|| 2−1) 2]

In this case, **l** ^x^ are samples that have been interpolated from produced and real data, and *λ*λ is a regularization parameter.

GANs' mathematical underpinnings offer a solid framework for comprehending how they work and how effective they are. The generator and discriminator are iteratively improved through the adversarial training process, which is framed as a minimax game. Convergence problems, however, present serious difficulties, which have prompted continued study and the creation of numerous methods to improve and maintain GAN training. Gaining an understanding of these mathematical foundations is crucial to developing and applying GANs in a variety of applications.

GAN TYPES

Since its debut, Generative Adversarial Networks (GANs) have undergone substantial development, resulting in the creation of numerous varieties of GANs with enhanced functionality and suited for certain applications. The most common kinds of GANs are examined in this section, along with their distinctive designs, benefits, and applications.

Figure 4. Types of GANs

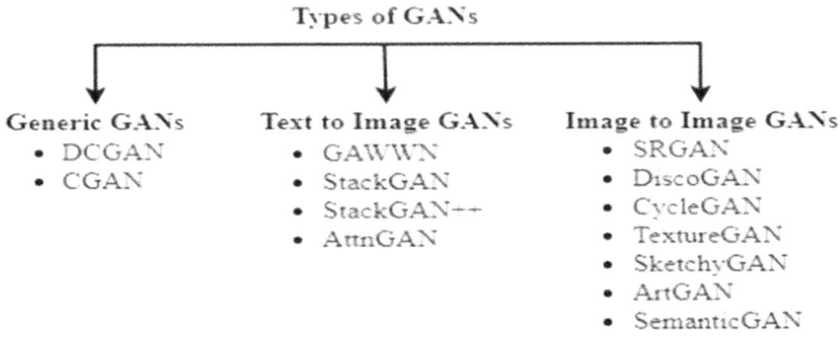

A Synopsis of Vanilla GANs

In 2014, Ian Goodfellow et al. presented the initial GAN model, sometimes known as Vanilla GAN. It provides the fundamental structure for all other GAN iterations.

Architecture Generator: A data sample is created from a random noise vector using a multi-layer perceptron (MLP).
Discriminator: An additional MLP that makes a distinction between created and real data samples.
Important Features
Use layers that are entirely integrated.
Use sigmoid activation in the output layer of the discriminator.
Use backpropagation and stochastic gradient descent to optimize.
Application
Applications Simple image creation activities.
Conceptual comprehension and instructional goals.

DCGANs, or Deep Convolutional GANs

Summary: In order to improve the quality of generated images, proposed Deep Convolutional GANs (DCGANs), which make use of convolutional neural networks (CNNs). (Nayak, M., et al. 2023).
Architecture Generator: Upsamples the input noise vector into a high-dimensional image by using deconvolutional (transposed convolution) layers.
Discriminator: Extracts hierarchical features for classification by downsampling the input image using neural layers.

Important Features

Convolutional layers are used in place of completely connected layers to improve the extraction of spatial features.

Employs batch normalization to make training stable.

Applies both the generator's and the discriminator's leaky ReLU activation.

Application

Creation of images of superior quality.

Super-resolution images and style transfer.

GANs With Conditions (cGANs)

Summary In 2014, Mirza and Osindero presented Conditional GANs (cGANs), which expand on the Vanilla GAN framework by adding more data (such class labels) to the discriminator and generator.

Architecture Generator: Takes as inputs a conditional vector (like the class label) and a noise vector.

Discriminator: To ascertain whether the data is authentic, it receives both the conditional vector and the data sample.

Important Features

Allows for the generation of data to be controlled by conditioning on auxiliary data.

Integrates conditional information using embedding or concatenation techniques.

Application

Translation from image to image.

Image-to-text conversion.

Data enrichment for particular classifications.

A Synopsis of Wasserstein GANs (WGANs)

In order to overcome training instability and mode collapse, presented Wasserstein GANs (WGANs), which provide a new loss function based on the Earth-Mover (Wasserstein) distance. (Nayak, M., et al. 2017).

Buildings like DCGANs or Vanilla GANs, but with changes made to the training protocols and loss function.

Important Features

Measures the difference between the produced and real data distributions using the Wasserstein distance.

Eliminates the sigmoid activation present in the output layer of the discriminator.

Imposes the Lipschitz restriction by applying weight clipping.
Applications
GANs are trained consistently on a range of tasks.
Enhanced image production with reduced artifacts.

A Synopsis of Least Squares GANs (LSGANs)

In order to address the vanishing gradient issue in the discriminator, (Mao et al. 2017) created Least Squares GANs (LSGANs), which alter the loss function to employ least squares errors.

Buildings
Like DCGANs, but with a loss function based on least squares.
Important Features
Provides the generator with smoother gradients by using a least squares loss.
Aids in producing photographs of excellent quality with fewer artifacts.
Uses
responsibilities involving the creation and restoration of images.
Applications needing smooth gradients and high-quality outputs.

A Synopsis of CycleGANs

(Zhu et al. 2017) presented CycleGANs, which guarantee that translating an image to another domain and return yields the original image, enabling unpaired image-to-image translation.

Buildings
Makes use of two discriminators and two generators.
Images from domain X are transformed into domain Y using generator G.
Images from domain Y are transformed into domain X using generator F.
Differentiate between translated and genuine images in domains X and Y using discriminators DX and DY, respectively. (Ather, D., et al. 2022).
Important Features
Introduces cycle consistency loss in order to guarantee consistency in the bidirectional transformation.
Does not need training data that is paired.
Uses
Transference of artistic styles.
Transformation of objects (e.g., changing horses become zebras).
Domain adaption across a range of industries.

A Synopsis of StyleGAN

2018 saw the introduction of StyleGAN by Karras et al., which aims to produce high-resolution, incredibly realistic images by enhancing control over the style and structure of generated images.

Buildings
Employs a unique generator design that modulates the intermediate layers dependent on style.
Incorporates noise inputs on several scales for fine-grained control.
Important Features
allows for different levels of fine-grained control on image characteristics.
Produces photos with excellent detail and resolution.
Uses
Creation of high-resolution images for media and entertainment.
Producing lifelike artwork and portraiture.

A Synopsis of BigGAN

In order to produce high-fidelity images, (Brock et al. 2018) proposed BigGAN, which scales up the GAN architecture and training procedure by utilizing big batch sizes and substantial computational resources.

Buildings
Use a discriminator and generator that are wider and deeper.
Includes self-attention mechanisms and class-conditional normalization.
Important Features
Increases the size of the training batch and the model to produce better-quality images.
Makes use of substantial datasets for training.
Uses
Producing varied and high-resolution photos for a range of uses.
Augmenting data for massive machine learning models.

Many variations of GANs have been developed as a result of its progress, each intended to solve a particular problem and enhance the original GAN architecture. These versions allow a wide range of applications, from domain adaptation and style transfer to high-quality image production using models as simple as Vanilla GANs and as complex as StyleGAN and BigGAN. Selecting the right model for a given job and developing the area of generative modeling require an understanding of the distinct architectures and features of different GANs. (Mayer, K. M., et al. 2011).

METHODS OF TRAINING FOR GANS

The process of training Generative Adversarial Networks (GANs) is difficult because of problems with mode collapse, instability, and hyperparameter sensitivity. The different methods and best practices created to enhance GAN training are covered in this section.

Appropriate Starting

Stable training depends on the weights in the generator and discriminator being initialized correctly. Gradients that explode or vanish due to poor initialization might cause instability in the training process.

Xavier Initialization: This technique helps preserve the variation of activations across layers and is frequently applied to layers with ReLU activations.
He Initialization: Fits for layers with ReLU activations; comparable to Xavier but with a somewhat different scaling factor.

Sorting by Batch

By normalizing each layer's input, batch normalization is used to stabilize the learning process. It lessens sensitivity to startup and speeds up training. Feizi, S., et al. (2020).

Generator: Every layer—aside from the output layer—is followed by batch normalization.
Discriminator: In order to prevent instability and "leaky" data reliance, batch normalization is frequently disregarded in the discriminator.

Gradual Penalties

A regularization component called a gradient penalty is introduced to the loss function in order to ensure that the discriminator's decision boundary is smooth.

Wasserstein GAN with Gradient Penalty, or WGAN-GP, enforces the Lipschitz constraint without weight clipping by adding a gradient penalty term to the WGAN loss:

L v =L D +λ E l ^ ~ p x ^ [(∑ ∇ x ^ D x ^) || 2 – 1) 2 L D = L D + λE x^~p x^ [(||∇ x ^ ^ D(x^)|| 2−1) 2]

In this case, l ^x^ are samples that have been interpolated from produced and real data, and λλ is a regularization parameter. Lewis, (S. M., et al. 2022).

Normalization of Spectral Data

By normalizing the spectral norm of every layer, spectral normalization regulates the Lipschitz constant of the discriminator. This method stabilizes GAN training without the need for weight clipping or gradient penalty. (Nayak, M., et al. 2023).

Smoothing Labels

Label smoothing entails giving the discriminator smoother labels rather than rigid labels. For instance, genuine photographs can be tagged with a value marginally less than 1 (e.g., 0.9), rather than being labeled as 1.

Benefits: Lessens the discriminator's confidence, assisting in preventing overfitting and giving the generator more useful gradients.
Injection of Noise
Regularizing the model and enhancing resilience can be achieved by directly injecting noise into the data samples or into the discriminator's inputs. Lewis, (S. M., et al. 2022).
Dropout: A technique used in the discriminator to prevent overfitting and introduce noise during training.
Gaussian Noise: Training can be stabilized by adding Gaussian noise to the discriminator's input.

Small-Scale Discrimination

By focusing on mini-batches rather than individual samples, mini-batch discrimination enables the discriminator to identify a lack of diversity in the generated samples.

Method: To reduce mode collapse, the discriminator analyzes a mini-batch of data and looks for patterns in the samples. (Feizi, S., et al. 2020).

Comparing Features

By altering the generator's objective to match the statistics of features taken from the discriminator's intermediate layer, feature matching stabilizes GAN training.

The generator's goal is to produce data in an intermediary layer of the discriminator that matches the mean and variance of the original data characteristics.

The Two-Time-Scale Update Rule (TTUR) modifies the discriminator and generator at varying rates of learning. In general, the discriminator learns faster than the generator or receives updates more frequently.

> **Advantages:** Promotes more steady training by assisting in the balancing of the generator and discriminator's learning rates.
> Example Sounds
> Adding noise to the input images and gradually annealing (reducing) it over training is known as instance noise.
> The goal is to smooth the decision border and keep the discriminator from growing overconfident too soon in the training process.

Perpendicular Regularization

The generator and discriminator weight matrices must be orthogonal in order for orthogonal regularization to work. This regularization technique aids in stabilizing training and preserving variability in the generated samples.

> Past Mean

By including a component in the loss function that penalizes the present parameters for deviating from their historical average, historical averaging is accomplished.

> **Implementation:** Adds a regularization term to the loss and monitors the network parameters' moving average.
> Mechanism of Self-Attention

When generating or discriminating data, the model can concentrate on pertinent portions of the input thanks to self-attention processes.

> **Applications:** Especially helpful for activities like creating complicated and large-scale images that need for an awareness of long-range dependencies. (Feizi, S., et al. 2020).

Gradual Expansion of GANs

According to Karras et al., progressive growing of GANs entails progressively raising the generated image resolution throughout training. (Saxena, D., et al. 2021).

Method: To enhance the resolution, begin with low-quality photos and gradually add layers to the discriminator and generator.
Advantages: Provides reliable training and produces high-quality images, especially for activities requiring high resolution.

Effective GAN training necessitates carefully weighing different strategies to deal with innate problems including instability, mode collapse, and hyperparameter sensitivity. Researchers and practitioners can accomplish more stable and effective training of GANs by utilizing approaches such as correct initialization, batch normalizing, gradient penalty, spectral normalization, label smoothing, noise injection, mini-batch discrimination, and feature matching. Furthermore, the capabilities and performance of GAN models are further improved by sophisticated techniques as TTUR, instance noise, orthogonal regularization, historical averaging, self-attention mechanisms, and progressive growing. Gaining an understanding of and utilizing these methods is necessary to effectively train GANs and realize their full potential across a range of applications.

EVALUATION METRICS FOR GANs

Generative Adversarial Networks (GANs) are difficult to evaluate since their performance cannot be readily measured with typical supervised learning criteria. Rather, a number of specific metrics have been created to evaluate the diversity and quality of the data that is generated. An overview of the most popular metrics for GAN evaluation is given in this section.

Figure 5. Evaluation metrics for GANs

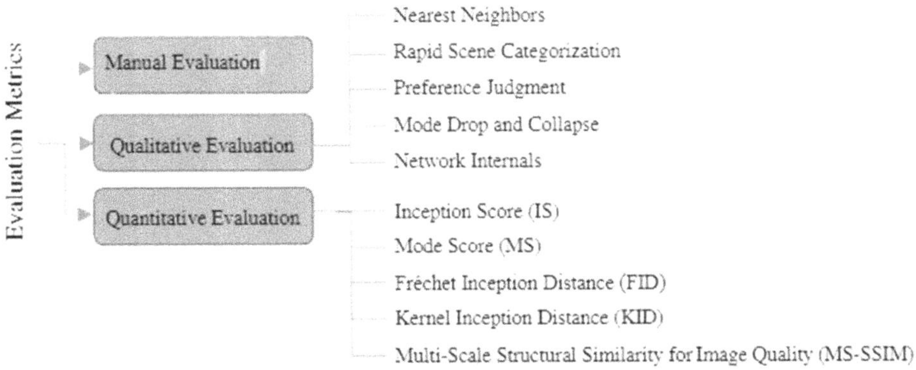

Inception Score (IS)

The output of a pre-trained Inception v3 network is used to determine the quality of generated images via the Inception Score.

Step 1: Use the GAN to create a collection of photos.
Step 2: To get the conditional label distribution p(y−x), run these photos through a trained Inception v3 network.
Step 3: Average the marginal label distribution *p* (*y* |) p(y|x) over all generated images to determine the marginal label distribution *p* (*y*) p(y).
Step4: Determine the KL divergence between *p* (*y* |) p(y|x) and *p* (*y*) p(y).

Finally, the Inception Score is provided by:

exp (*E x* [*D* N *L* (*p* (*y* | *x*)|^ *p* (*y*)]) exp(E * x **D KL * (p(y) * x) * p(y))])

Interpretation
Great Score: Denotes a great degree of diversity and quality in the photographs produced.
Low Score: Indicates that the generated photos are of low quality or diversity.
Benefits
Gives a single score that incorporates diversity and visual quality.
Quite popular and simple to use. (Saxena, D., et al. 2021).
Restrictions
Sensitive to the pre-trained model selection.
Does not quantify the resemblance to actual data directly.

Frechet Inception Distance (FID)

The distributions of genuine and artificially generated images using characteristics taken from a pre-trained Inception v3 network are compared using the Frechet Inception Distance.

Calculation

Step 1: Gather a set of authentic photographs and create a set of images using the GAN.
Step 2: To extract feature representations, run both sets of images through an Inception v3 network that has already been trained.
Step 3: Determine the characteristics' mean and covariance for both real and (μr, Σr) and generated images (μg, Σg).
Step 4: Determine the two multivariate Gaussians' Frechet distance from each other:

$$FID=||\mu r-\mu g||^2+\mathrm{Tr}(\Sigma r+\Sigma g-2(\Sigma r\Sigma g)^{1/2})$$

Interpretation
A low FID suggests high-quality and diversified created images since it shows that the distributions of genuine and generated images are similar.
High FID: Indicates that the produced and real picture distributions differ significantly from one another.
Benefits
Evaluates the diversity and quality of the generated photos.
Stronger than IS against noise and mode collapse. (Saxena, D., et al. 2021).
Restrictions
computationally demanding because of the calculations for the covariance matrix.
Sensitive to the quantity of samples and the selection of the pre-trained model.

Precision and Recall Metrics for GANs

These metrics assess the quality and diversity of generated images in a manner similar to how they are used in classification tasks.

Calculation

This metric measures the fraction of generated images that are near to the real data manifold.

Recall: This metric measures the fraction of the real data manifold covered by the generated images.

Interpretation

High Precision indicates that generated images are high quality and closely resemble real images.

High Recall: This metric indicates that generated images are diverse and adequately cover the real data distribution.

Benefits

Identifies quality and diversity independently. Offers more granular insight into the performance of the GAN.

Limitations

This method necessitates the careful definition of distance metrics and thresholds.

Kernel Inception Distance (KID)

Similar to FID, Kernel Inception Distance compares the distributions of produced and real images using polynomial kernels. (Zhang, W., et al. 2018).

Calculation

Step 1: Use a pre-trained Inception v3 network to extract feature representations from generated and actual images.

Calculate the polynomial kernel that connects the characteristics of the produced and real images in step two.

Step 3: Determine the two sets of features' squared Maximum Mean Discrepancy (MMD).

$$KID=(1/m(m-1))\sum i\neq jk(xi,xj)+(1/n(n-1))\sum i\neq jk(yi,yj)-(2/mn)\sum i,jk(xi,yj)$$

where *xi* and *xj* are features of real images, *yi* and *yj* are features of generated images, and *k* is the polynomial kernel.

Interpretation

Low KID: Shows that the produced and real picture distributions are comparable.

High KID: Indicates that the produced and real image distributions differ significantly from one another.

Benefits

It is an unbiased estimator that requires fewer samples to compute.

Robust compared to FID at small sample quantities.

Restrictions

Sensitive to the kernel and its parameter selection.

Computationally heavy. (Saxena, D., et al. 2021).

Multi-Scale Structural Similarity (MS-SSIM)

Luminance, contrast, and structure are taken into account when evaluating the similarity of image pairings using multi-scale structural similarity.

Calculation

First, create image pairings and calculate the SSIM index at various scales.

Step 2: Take the average of all pairings' and scales' SSIM indices.

Interpretation

High MS-SSIM: A sign of structural similarity between the produced and original pictures.

Low MS-SSIM: Indicates that artificial and genuine images differ significantly in their structural makeup.

Benefits

provides a more thorough similarity measure by taking into account image quality at various scales.

Sensitive to structural details, which makes it appropriate for evaluating the quality of images.

Restrictions

Does not quantify diversity directly.

Sensitive to the selection of weights and scales.

Mode Score

By taking into account the diversity of created images explicitly, the Mode Score expands upon the Inception Score.

Calculation

This is similar to the Inception Score, except it adds a penalty term for mode collapse.

Mode Score is equal to $\exp(E_x)\ [D_{KL}\ (p(y|x)||p(y))]-D_{KL}\ (p(y)||q(y))))$

And the genuine label distribution is represented by $q\ (y)$ q(y).

Interpretation

High Mode Score: Denotes a varied range of high-quality created photos.

Low Mode Score: Indicates a lack of diversity or poor quality.

Benefits

clearly penalizes mode collapse, resulting in a more impartial evaluation of diversity and quality.

Restrictions

computationally demanding and contingent on the pre-trained model selection.

Specialized criteria that take into consideration the diversity and quality of the generated data are needed to evaluate GANs. A variety of metrics provide distinct insights into the performance of GANs, including Inception Score (IS), Frechet Inception Distance (FID), Precision and Recall, Kernel Inception Distance (KID), Multi-Scale Structural Similarity (MS-SSIM), and Mode Score. Every metric has benefits and drawbacks, and in order to provide a thorough assessment, a mix of these metrics is frequently employed. Comprehending these indicators is crucial for evaluating and enhancing GAN performance across a range of applications. (Zhang, W., et al. 2018).

ADVANCED TOPICS IN GANs

Since their debut, Generative Adversarial Networks (GANs) have undergone substantial development. Beyond the fundamental architecture and training methods, a number of cutting-edge subjects have surfaced that expand on the capabilities of GANs. Conditional GANs, semi-supervised learning, disentangled representations, transfer learning, self-supervised learning, and applications across multiple domains are some of the advanced topics that are covered in this part.

Conditional GANs (cGANs)

By conditioning the generator and discriminator on extra information, such as class labels or data properties, conditional GANs expand on the basic GAN architecture. More control over the output that is generated is made possible by this. Nayak, M., et al. (2023).

Important Ideas

Conditional Information: Both the discriminator and the generator receive additional information y (such as class labels).

The generator's objective function is to generate samples that match the specified criterion y and look realistic.

Uses

Image-to-Image Translation: This includes tasks like day-to-night conversions or the conversion of sketches to photographs.

Text-to-Image Synthesis: Producing visual representations from written descriptions.

Disentangled Representations

The goal of disentangled representations is to isolate the several variables that cause variance in the data, such as identity, lighting, and posture in pictures. To learn these disentangled elements, GANs can be applied.

Important Ideas

Latent Space: The input space of the generator is organized to reflect distinct components of variation independently.

Control: Certain elements of the generated output can be tailored by adjusting specified latent vector dimensions.

Methods

InfoGAN: Encourages the generator to develop interpretable and disentangled representations by introducing a mutual information term.

To enforce disentanglement, Beta-VAE combines variational autoencoders with GANs.

Uses

Style Transfer: To properly use style transfer, keep content and style separate.

The creation of controlled avatars for Human-Computer Interaction.

Using GANs to Transfer Learning

A pre-trained GAN model is adapted to a new but similar task or dataset through the process of transfer learning. This is especially helpful when there are limited computational resources or when the new dataset is tiny. (Zhang, W., et al. 2018).

Important Ideas

Pre-training: A sizable and varied dataset is used to train a GAN initially.

Fine-tuning: After being pre-trained, the GAN is adjusted using the target dataset, usually at a lower learning rate.

Methods

Layer Freezing: While higher levels of the GAN are being adjusted, some layers—typically the lower layers—are left unchanged.

Domain Adaptation: To produce data that matches the target domain distribution, the GAN is modified.

Uses

Medical Imaging: Using sparse data to modify previously trained GANs for particular medical imaging applications.

Art Generation: Fine-tuning GANs trained on general image datasets to create art in specific styles. (Ather, D., et al. 2022).

Self-Supervised Learning

Self-supervised learning makes use of unlabeled data by using pretext tasks to generate pseudo-labels. Self-supervised learning is a useful tool for GANs to enhance their generated outputs and representations.

Important Ideas

Pretext Tasks: These are tasks like jigsaw puzzles and rotation predictions that don't need manual labeling.

Learning valuable characteristics from the data itself without explicit labels is known as representation learning. (Y. K. R., et al. 2017).

Methods

RotNet: Developing a self-supervised GAN that can identify photos that have been rotated.

Contrastive learning involves teaching the GAN to discern between picture pairs that are similar and those that are not.

Uses

Enhancing the quality of learned features for use in subsequent processes such as segmentation or classification is known as feature learning.

Data efficiency: Lessening the requirement for huge datasets with labels.

Applications across a Range of Subjects

GANs have been used in many different fields, proving their adaptability and potential influence.

Computer Vision

Image Super-Resolution: Increasing an image's resolution.

Image inpainting: Replacing areas of images that are absent.

Object detection: Using augmented data to improve object detection models.

Natural Language Interpretation
Creating visuals from written descriptions is known as text-to-image synthesis.
Enhancing the caliber of conversations produced by conversational agents is known as dialogue generation.
Health Imaging
MRI Reconstruction: Using undersampled data, high-quality MRI pictures are reconstructed. (Zhang, W., et al. 2018).
Disease Diagnosis: Adding artificial medical imagery to improve diagnosis models.
Audio Reproduction Speech Synthesis: Producing artificial speech of superior quality.
Music generation is the process of writing new songs using patterns that are learnt.

FUTURE DIRECTIONS FOR GANs

Generative Adversarial Networks, or GANs, are widely applicable and have transformed the area of generative modeling. There are still a lot of obstacles to overcome and plenty of room for more study and advancement. This section examines a number of exciting potential future avenues for GANs, including as utilizing multi-modal data, addressing ethical problems, boosting interpretability, and strengthening training stability, among other things. Y. K. R., et al. (2017).

Figure 6. Future of GANs

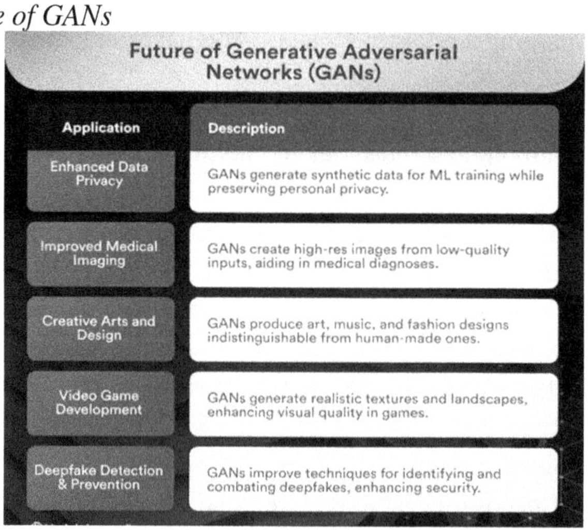

The future of GANs involves several promising directions:

Enhancing Training Stability: Creating sophisticated algorithms and methods, like adaptive learning rates, cutting-edge regularization strategies, and improved loss functions, to guarantee steady and dependable convergence.
Improving Interpretability: GANs can be made more transparent and intelligible by using advanced visualization tools, disentangled representations, and explainable AI (XAI) techniques.
Handling Ethical Issues: Developing strong deepfake detection methods, defining moral standards, and encouraging openness and responsibility in the application of GAN-generated material.
Utilizing Multi-Modal Data: To facilitate the integration and production of complex outputs from various data kinds, cross-modal GAN development, unified representation learning, and multi-modal dataset curation are employed.
Expanding Applications: Increasing the effect and usefulness of GANs by using them in new fields like education, tailored medicine, scientific research, and artistic expression.
Federated Learning: Integrating GANs with federated learning to allow distributed training, scalability between decentralized devices, and privacy-preserving generative modeling.
Adaptive and Self-Improving GANs: Using adaptive architectures, continuous learning frameworks, and meta-learning will enable GANs to learn and adjust in real-time to new tasks and data.

With major improvements anticipated in training stability, interpretability, ethical issues, multi-modal data integration, application growth, federated learning, and adaptive capabilities, the future of GANs is promising and could be revolutionary. GANs are expected to grow in strength and adaptability as scientists and industry professionals innovate and tackle present issues, opening up new avenues for research and use in a variety of sectors. To fully utilize GANs for societal good, responsible development and implementation that is informed by moral principles and openness will be necessary.

CONCLUSION

In the realm of artificial intelligence, Generative Adversarial Networks (GANs) have become a revolutionary technology because of their unparalleled ability to generate realistic data in a variety of modalities, including text, audio, and images. A generator and a discriminator operating in a competitive framework make up the basic architecture of GANs, which has shown to be a potent method for producing high-quality synthetic data that closely resembles real-world examples. Notwithstanding their achievements, GANs still have a number of drawbacks, such as instability in training, mode collapse, and issues assessing the caliber of generated data. Numerous modifications and enhancements, including conditional GANs, Wasserstein GANs, and semi-supervised learning approaches, have been made possible by advances in GAN research. These innovations help to overcome some of these difficulties and increase the potential of GANs. With an eye toward the future, a number of encouraging paths are in place to further improve the efficiency and usefulness of GANs. Important areas of ongoing research include developing robust detection methods and ethical frameworks to address ethical concerns, leveraging multi-modal data for richer and more diverse outputs, improving interpretability through explainable AI techniques, and improving training stability through advanced algorithms. Furthermore, there are a lot of innovative prospects in constructing adaptive, self-improving GANs, incorporating federated learning for modeling that preserves privacy, and extending GAN applications into other domains including scientific research, healthcare, and education. As GANs develop further, they have the potential to significantly advance a number of fields, providing instruments for practical applications that can change industries and enhance lives as well as tools for scientific research and creative expression. To maximize the advantages of GANs while minimizing potential hazards, however, responsible development and deployment—guided by ethical considerations and a commitment to transparency—will be necessary. GANs are expected to be crucial to the development of artificial intelligence in the future with further study and cooperation.

FUTURE SCOPE

The foundations of Generative Adversarial Networks (GANs) have a wide-ranging and promising future with applications in many different sectors. The field of generative modeling has undergone a radical transformation thanks to GANs, which are made up of a discriminator and a generator playing a minimax game. Here are some predictions for the future:

Improved Image and Video Generation: GANs will keep getting better at producing realistic, high-quality photos and videos. This encompasses augmented reality, virtual reality, and entertainment applications, where the development of lifelike content is essential.

Medical Imaging: GANs have the potential to improve medical imaging methods by allowing low-resolution data to be converted into high-resolution images. This can help with more precise therapy planning and diagnosis.
Data Augmentation: GANs will be important in this regard, especially in situations where there is a lack of labeled data. In domains where obtaining different datasets is difficult, such as autonomous driving, this is crucial for training effective machine learning models.
Creative Arts: Through the creation of art, music, and literature, GANs are poised to transform creative fields. They can help artists and makers by offering fresh resources for ingenuity and originality.
Scientific Research: GANs are useful in subjects like physics, chemistry, and climate science because they can model complicated physical systems and processes.
Privacy and Security: GANs in cybersecurity can provide synthetic data that protects privacy, allowing for safe data analysis and sharing. By imitating potential hazards, they can also be utilized to detect and neutralize hostile activity.
Better Training Methods: In order to create more durable and dependable models, future research in GAN training is probably going to concentrate on resolving existing issues with mode collapse and instability.

The underlying principles of GANs will keep developing, spurring innovation and revolutionizing sectors by allowing machines to produce and comprehend complicated data with previously unheard-of levels of sophistication

REFERENCES

Alqahtani, H., Kavakli-Thorne, M., & Kumar, G. (2021, March). Applications of generative adversarial networks: An updated review. *Archives of Computational Methods in Engineering*, 28(2), 525–552. DOI: 10.1007/s11831-019-09388-y

Ather, D., Madan, S., Nayak, M., Tripathi, R., Kant, R., Kshatri, S. S., & Jain, R. (2022). Selection of smart manure composition for smart farming using artificial intelligence technique. *Journal of Food Quality*, 2022(1), 4351825.

Bhat, J. (2005, May). Concerns of new technology based industries—The case of nanotechnology. *Technovation*, 25(5), 457–462. DOI: 10.1016/j.technovation.2003.09.001

Cho, Y. K. R.. (•••). Sub-10 nanometer feature size in silicon using thermal scanning probe lithography. *ACS Nano*, 11, 11890–11897. PMID: 29083870

Cho, Y. K. R.. (•••). Sub-10 nanometer feature size in silicon using thermal scanning probe lithography. *ACS Nano*, 11, 11890–11897. PMID: 29083870

Fan, P., Gao, J., Mao, H., Geng, Y., Yan, Y., Wang, Y., Goel, S., & Luo, X. (2022, January 29). Scanning probe lithography: State-of-the-art and future perspectives. *Micromachines*, 13(2), 228. DOI: 10.3390/mi13020228 PMID: 35208352

Fan, P., Gao, J., Mao, H., Geng, Y., Yan, Y., Wang, Y., Goel, S., & Luo, X. (2022, January 29). Scanning probe lithography: State-of-the-art and future perspectives. *Micromachines*, 13(2), 228. DOI: 10.3390/mi13020228 PMID: 35208352

Generative adversarial networks: An overview of theoretical model, evaluation metrics, and recent developments. Retrieved from https://www.researchgate.net/figure/Categorization-of-GAN-evaluation-metrics_fig6_341699736

Intelligent Analytics for Industry 4.0 Applications.. CRC Press.

Le-The, H., Küchler, C., van den Berg, A., Bodenschatz, E., Lohse, D., & Krug, D. (2021, April 02). Fabrication of freestanding Pt nanowires for use as thermal anemometry probes in turbulence measurements. *Microsystems & Nanoengineering*, 7(1), 28. DOI: 10.1038/s41378-021-00255-0 PMID: 34567742

Le-The, H., Küchler, C., van den Berg, A., Bodenschatz, E., Lohse, D., & Krug, D. (2021, April 02). Fabrication of freestanding Pt nanowires for use as thermal anemometry probes in turbulence measurements. *Microsystems & Nanoengineering*, 7(1), 28. DOI: 10.1038/s41378-021-00255-0 PMID: 34567742

Lewis, S. M., DeRose, G. A., Alty, H. R., Hunt, M. S., Lee, N., Mann, J. A., Grindell, R., Wertheim, A., De Rose, L., Fernandez, A., Muryn, C. A., Whitehead, G. F. S., Timco, G. A., Scherer, A., & Winpenny, R. E. P. (2022, August). Tuning the performance of negative tone electron beam resists for the next generation lithography. *Advanced Functional Materials*, 32(32), 2202710. DOI: 10.1002/adfm.202202710

Mayer, K. M., & Hafner, J. H. (2011, June 08). Localized surface plasmon resonance sensors. *Chemical Reviews*, 111(6), 3828–3857. DOI: 10.1021/cr100313v PMID: 21648956

Nayak, M. Prelims of R. *International Journal of Technical Research and Applications, 5*, 35–37. www.ijtra.com

Nayak, M., & Barman, A. A real-time cloud-based healthcare monitoring system. In S. Kshatri, K. Thakur, M. Khan, D. Singh, & G. Sinha, *Computational Intelligence and Applications for Pandemics and Healthcare* (pp. 229–247). IGI Global. DOI: 10.4018/978-1-7998-9831-3.ch011

Nayak, M., & Dass, A. K. GSM and Arduino based smart home safety and security system. *Zenodo*. DOI: 10.5281/zenodo.7610756

Nayak, M., & Dass, A. K. GSM and Arduino based smart home safety and security system. *Recent Trends in Information Technology and Its Application, 6*, 1–6. DOI: 10.5281/zenodo.7610756

Nayak, M., Doss, A. K., & Kshatri, S. S. An AI-based efficient model for the classification of traffic signals using the convolutional neural network. In *Building Secure Business Models Through Blockchain Technology* (pp. 20–35). DOI: 10.4018/978-1-6684-7808-0.ch002

Nayak, M., & Narain, B. Predicting dynamic product price by online analysis: Modified K-means. In *Springer*. DOI: 10.1007/978-981-15-2449-3_1

Nayak, M., & Panda, S. *Online grocery recommendation system implementation using Python Flask and machine learning*. Retrieved from https://zenodo.org/records/10715185

Reyntjens, S., & Puers, R. (2001, July 01). A review of focused ion beam applications in microsystem technology. *Journal of Micromechanics and Microengineering*, 11(4), 287–300. DOI: 10.1088/0960-1317/11/4/301

Saifullah, M. S., Asbahi, M., Neo, D. C. J., Mahfoud, Z., Tan, H. R., Ha, S. T., Dwivedi, N., Dutta, T., bin Dolmanan, S., Aabdin, Z., Bosman, M., Ganesan, R., Tripathy, S., Hasko, D. G., & Valiyaveettil, S. (2022, September 28). Patterning at the resolution limit of commercial electron beam lithography. *Nano Letters*, 22(18), 7432–7440. DOI: 10.1021/acs.nanolett.2c02339 PMID: 36069429

Saxena, D., & Cao, J. (•••). Generative adversarial networks: Challenges, solutions, and future directions. *ACM Computing Surveys*, 54, 63. Advance online publication. DOI: 10.1145/3446374

Shehadeh, M., Ebrahimi, N., & Ochigbo, A. (2012). Predicting the type of nano-structure using data mining techniques and multinomial logistic regression. *Procedia Computer Science*, 12, 392–397. DOI: 10.1016/j.procs.2012.09.092

Zhang, W. (2018, June). Generative adversarial nets for information retrieval: Fundamentals and advances. In The 41st International ACM SIGIR Conference on Research & Development in Information Retrieval (pp. 1375-1378).

Chapter 8
Building Your First GAN Model

Priya Parkhi
Ramdeobaba University, Nagpur, India

Samiksha Harjal
Ramdeobaba University, Nagpur, India

Bhagyashree Hambarde
https://orcid.org/0000-0002-7331-5164
St. Vincent Pallotti College of Engineering and Technology, Nagpur, India

Himesh Ganwani
Ramdeobaba University, Nagpur, India

Rupali Sachin Vairagade
Shah and Anchor Kutchhi Engineering College, Mumbai, India

ABSTRACT

Generative Adversarial Networks (GANs), introduced by Ian Goodfellow in 2014, have revolutionized machine learning by creating new data. GANs consist of two parts: the Generator, which creates data samples from random noise, and the Discriminator, which distinguishes real from fake data. These networks compete, with the Generator improving its fakes and the Discriminator enhancing its detection. This competition drives both networks to improve. Training GANs can be challenging due to issues like instability and mode collapse. Advanced versions like Conditional GANs (c GANs) and Wasserstein GANs (WGANs) address these problems. This chapter introduces GANs, explaining their components, adversarial training, and practical implementation using the MNIST dataset. It includes guidance on tools, dataset understanding, architecture design, and visualization techniques

DOI: 10.4018/979-8-3693-7152-7.ch008

for evaluating generated images.

INTRODUCTION TO GENERATIVE ADVERSARIAL NETWORKS (GANS)

Generative Adversarial Networks (GANs) an Ian Goodfellow and his team introduced a specific type of neural network in 2014. These networks are particularly powerful for unsupervised learning, where the objective is to identify trends within the data. without labeled examples. GANs can create new information that resembles the data used during their training, such as realistic images, sounds, or text.

To understand GANs, it's helpful to first know a bit about Convolutional Neural Networks (CNNs). CNNs are designed to classify images by analyzing them pixel by pixel. When you feed an image to a CNN, it processes the image through its hidden layers and determines what the image represents, such as whether it contains a dog or a cat. This makes CNNs very effective for tasks like image classification.

GANs have revolutionized artificial intelligence by enabling models to generate data with high realism. This chapter aims to introduce GANs, explain their significance in AI, and provide a step-by-step guide to building your first GAN model.

GANs, however, take a different approach. Instead of just classifying data, they generate new data. The core idea behind GANs involves pair of neural networks that compete with each other. One network, called the generator, creates new data. The other network, called the discriminator, evaluates this data against real data. The generator aims to produce data that closely resembles real data, while the discriminator's role is to determine if the data is authentic or artificially created. Through this competition, both networks improve over time, resulting in highly realistic generated data.

Because of this unique structure, GANs have become very popular for generating high-quality images, sounds, and even text. They are used in various applications, from creating realistic animations to enhancing data with synthetic examples. GANs represent a cutting-edge approach in deep learning, capable of producing new data that closely mimics the original dataset.

What are GANs?

A Generative Adversarial Network (GAN) consists of two distinct neural networks: the Generator (G) and the Discriminator (D). These two models function within a game-theoretic framework, where they engage in a continuous competition

Generator: The Generator in a GAN is like an artist creating forgeries. It receives random noise as input and aims to produce data such as images, sounds, or text—that is realistic enough to fool the Discriminator. Structurally, it's akin to an inverse Convolutional Neural Network (CNN). Instead of taking an image and outputting a label, the Generator starts with random noise and generates data that should look like it came from the real dataset.

Discriminator: The Discriminator is responsible for distinguishing real data from the synthetic data generated by the Generator. It's essentially a binary classifier, trained on real data to learn what genuine data looks like. Unlike traditional CNNs that output probabilities for various classes, the Discriminator in a GAN has a simpler task: outputting a probability that the input data is real (1) or fake (0). It helps the GAN by providing feedback to the Generator when the Generator creates low-quality data that the Discriminator easily identifies as fake, the Generator adjusts its parameters to improve.

Figure 1. Overview of GAN structure

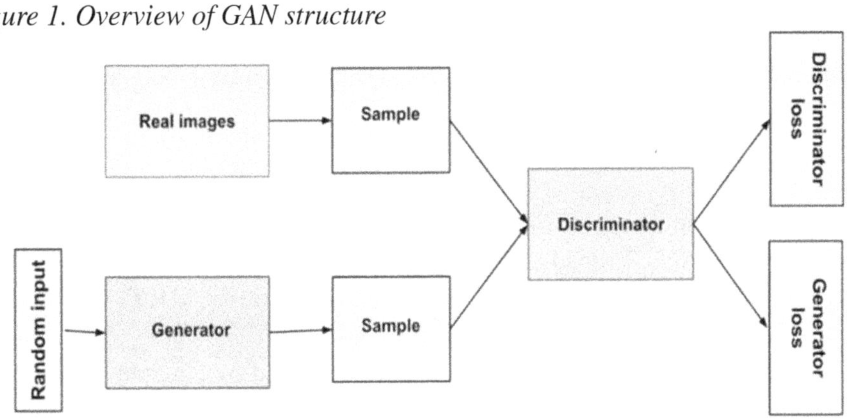

How They Work Together: GANs work through an adversarial process in which the Generator and Discriminator challenge each other in a game-like interaction. The Generator aims to create data that closely resembles real samples, making it increasingly difficult for the Discriminator to detect any differences. At the same time, the Discriminator progressively enhances its ability to differentiate between real and generated data. This constant interplay forces both networks to improve over time.(Figure 1)

Types of GANs

1. Vanilla GAN

The Vanilla GAN is the most straightforward form of GAN. It consists of a Generator and a Discriminator, both designed as basic multi-layer perceptrons. The primary objective in a Vanilla GAN is to optimize a mathematical function through the use of stochastic gradient descent. Despite its simplicity, the Vanilla GAN laid the foundation for more advanced GAN variants.

2. Conditional GAN (CGAN)

A Conditional GAN (CGAN) introduces additional parameters into the model. It incorporates extra information, or conditions, into both the Generator and the Discriminator. For instance, an extra parameter 'y' could be introduced to the Generator to generate data that aligns with particular conditions.The Discriminator also receives these labels as inputs to help distinguish between real and fake data more accurately. This conditioning allows CGANs to generate data based on specific requirements, making them effective for tasks where context or labels are important.

3. Deep Convolutional GAN (DCGAN)

The Deep Convolutional GAN (DCGAN) is one of the most popular and successful kinds of GANs. Instead of using simple multi-layer perceptrons, DCGANs use Convolutional Neural Networks (ConvNets). Key features of DCGANs include:

- **No Max Pooling:** Max pooling layers are replaced by convolutional stride, which helps in preserving spatial hierarchies in images.
- **No Fully Connected Layers:** Fully connected layers are avoided to maintain the structure and integrity of image data.

DCGANs are particularly effective in generating high-quality images due to their use of ConvNets, which are well-suited for image processing tasks.

4. Laplacian Pyramid GAN (LAPGAN)

The Laplacian Pyramid GAN (LAPGAN) leverages the Laplacian pyramid, a structured image representation that is both linear and invertible. This model employs multiple Generator and Discriminator networks at different levels of the pyramid. Initially, the image undergoes down-sampling at each level before being progressively up-scaled in a reverse process. During this up-scaling phase, noise is introduced by the Conditional GAN at every stage until the image is reconstructed to its original dimensions. The multi-level refinement approach of LAPGANs enables the generation of highly detailed and realistic images.

5. Super Resolution GAN (SRGAN)

Super Resolution GAN (SRGAN) focuses on enhancing the resolution of images.It employs a deep neural network in conjunction with an adversarial network to generate images with higher resolution from low-resolution inputs. SRGANs are particularly useful for applications that require detailed and clear images, as they can effectively upscale low-resolution images while minimizing errors and preserving details.

Each of these GAN types brings unique improvements and capabilities, making GANs versatile tools in the field of generative modeling. From basic image generation to high-resolution enhancements and context-specific data generation, GANs continue to evolve and find new applications across various domains.

Table 1. Types of GAN

Type of GAN	Advantages	Limitations	Applications
Vanilla GAN	Simple and easy to implement	Prone to mode collapse	Basic image generation
Conditional GAN (CGAN)	Allows for controlled data generation	Requires labeled data	Image-to-image translation, text-to-image synthesis
Deep Convolutional GAN (DCGAN)	Generates high-quality images	Struggles with diversity in generated samples	Realistic face generation, artwork synthesis
Super Resolution GAN (SRGAN)	Enhances image resolution effectively	Computationally expensive	Satellite imaging, medical imaging enhancement

How GANs Work

Generative Adversarial Networks (GANs) operate through a unique training process that can be likened to a two-player game, involving a Generator and a Discriminator. Here's a detailed look at how GANs work:

Initialization:
Create Two Neural Networks:
Generator (G): This network is tasked with generating new data samples that mimic real data.

Discriminator (D): This network acts as a critic that distinguishes between actual data (from training set) and synthetic data (generated by Generator).

The Training Process

GANs are trained through a process where the Generator and Discriminator are updated in successive iterations. This adversarial process allows both networks to continuously improve.

Step 1: Generator Training

Initial Generation: The Generator starts by taking a random noise vector, filled with arbitrary values, is used as an initial input to generate new data samples, such as images.

Transformation: Utilizing its internal layers and learned patterns, the Generator converts the noise vector into a new data sample. Initially, these generated samples are of poor quality.

Step 2: Discriminator Training

Receiving Inputs: The Discriminator receives two types of inputs:

- Real data samples from the training dataset.
- Fake data samples generated by the Generator.

Classification: The Discriminator analyzes each input and outputs a probability score between 0 and 1:

- A score close to 1 indicates the data is likely real.
- A score close to 0 suggests the data is fake.

Adversarial Training Loop: This loop consists of iteratively training the Discriminator and the Generator.

Training the Discriminator:

1. **Real Data Analysis:** The Discriminator is trained using a batch of genuine data from the training dataset.
2. **Fake Data Analysis:** Simultaneously, it is trained on a batch of synthetic data generated by the Generator.
3. **Loss Function:** The Discriminator's loss function usually involves binary cross-entropy loss, aiming to increase the likelihood of accurately distinguishing between genuine and synthetic samples.

Training the Generator:

1. **Creating Data:** The Generator produces synthetic data by processing random noise as input.
2. **Fooling the Discriminator:** The Generator aims to create data that the Discriminator will incorrectly identify as genuine.
3. **Loss Function:** The Generator's loss is derived from the Discriminator's output, aiming To reduce the likelihood that the Discriminator accurately classifies the generated data as fake.

Figure 2. Architecture of GANs

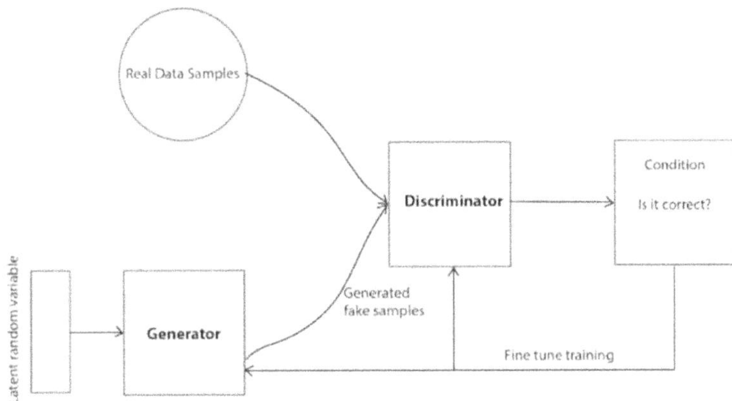

Iterative Improvement:

Discriminator's Reward and Penalty:

- When the Discriminator correctly identifies real data as real and fake data as fake, it receives a small reward.
- If the Discriminator misclassifies fake data as real, it gets a penalty, encouraging it to improve its discrimination abilities.

Generator's Feedback:

- When the Generator's fake data is classified as real by the Discriminator, the Generator receives positive feedback, helping it to refine its data generation process.
- Conversely, if the fake data is correctly identified as fake, the Generator receives feedback to improve its data generation to make it more realistic.

Continuous Refinement:

- The Generator and Discriminator continuously refine their abilities through this adversarial process.

- The Generator improves its ability to create realistic data, making it progressively harder for the Discriminator to differentiate between authentic and synthetic data.
- The Discriminator enhances its skill in detecting subtle distinctions between genuine and generated data.

Goal of Training:

- The ultimate goal is for the Generator to become so skilled that the Discriminator can no longer reliably differentiate between real and fake data.
- At this point, the Generator is considered well-trained and can produce high-quality, realistic data samples.

GANs often suffer from training challenges such as mode collapse, vanishing gradients, and instability. Mode collapse occurs when the Generator produces limited variations of data instead of a diverse set. Vanishing gradients hinder training when the Discriminator becomes too strong, making it difficult for the Generator to learn. To address these issues, techniques like Wasserstein loss, spectral normalization, and batch normalization are employed to stabilize training and encourage diversity in outputs.

In summary, GANs work through an ongoing competition between Generator and Discriminator, where each network constantly learns and improves from the feedback provided by the other. This dynamic interplay results in the Generator producing increasingly realistic data, which is the essence of how GANs function.

LITERATURE SURVEY

The literature review has been structured into key themes:

1. **Early Challenges in GANs** – Discusses issues such as training instability and mode collapse.
2. **Recent Advances** – Covers novel techniques like progressive growing GANs and hybrid GAN architectures.
3. **Applications** – Highlights GANs in fields like medical imaging, climate modeling, and finance.
4. **Future Research Directions** – Identifies gaps such as the need for improved evaluation metrics and ethical concerns.

Each cited paper now includes a summary of its contributions to the GAN field, along with a discussion of its limitations and potential areas for future research.

The paper offers a comprehensive review of the challenges encountered by Generative Adversarial Networks (GANs), issues like mode collapse, failure to converge, and instability during training, while proposing solutions and future directions for improvement. Its primary goal is to provide an organized summary of various strategies for designing and optimizing GANs, targeting new researchers in the field and addressing the lack of existing surveys systematically covering these developments. The study identifies key issues in GAN design and optimization, proposing a classification system to organize solutions based on these issues. It examines various GAN types developed for diverse applications, providing detailed descriptions and analyses of how they address the challenges. Additionally, the research explores novel solutions and the extensive application of GANs in different research areas, offering insights into their real-world use. Finally, the paper outlines promising future research directions for the rapidly evolving GAN field, highlighting potential advancements and improvements in design and optimization strategies. (Saxena et al., 2021)

Generative Adversarial Networks, commonly known as GANs, are unique computer programs that can learn to generate new content, such as images or music. What sets GANs apart from other similar technologies is the way they improve their creations. Instead of just trying to be as efficient as possible, GANs engage in a kind of game to enhance their output. In this game, there are two parts: one that creates the content and another that evaluates it. The creator tries to make the most realistic content possible, while the evaluator tries to determine if the content is real or fake. Through this back-and-forth process, both parts get better at their tasks, resulting in highly detailed and realistic outputs.Although GANs have proven to be incredibly useful in various research areas, they are still quite tricky to work with. Scientists are continually studying them to understand their complexities and to improve their ability to create even more realistic and diverse content. (Goodfellow, 2020)

Generative adversarial networks (GANs) represent a category of machine learning models that can understand and represent complex data patterns without needing a lot of labeled examples. They consist of two networks: one generates new data, while the other tries to tell if the data is real or generated. GANs are versatile and can be used for a variety of tasks, including creating new images, editing existing images, transferring artistic styles between images, improving image quality, and even classifying images.The main aim of the review article is to introduce GANs to people working in signal processing, explaining how GANs are built, how they are trained, and the challenges that come with using them.The article covers different techniques for training and constructing GANs, stressing the importance of understanding the theory behind them to apply GANs effectively in various practical scenarios.Even though there have been significant advancements in GAN research, there are still many theoretical and practical issues that need to be resolved. This means that researchers are continuously working to enhance the performance and stability of GANs for different signal processing tasks.By giving a comprehensive overview of GANs and their applications in signal processing, the article aims to connect the theoretical aspects of GANs with their practical uses. This helps researchers to better understand and utilize GANs in their work. (Creswell et al., 2018)

The authors conducted a systematic review of recent studies on Generative Adversarial Network (GAN) models and their various applications. They searched through databases like Embase (Scopus), Web of Science (WoS), and PubMed, finding a total of 2084 documents related to GAN. After a thorough screening process, they selected 52 records for their final review.The study found that GANs are being used in several fields including 3D object generation, medicine, pandemic-related applications, image processing, face detection, texture transfer, and traffic control. They noted that research on GANs was limited before 2016 but has significantly increased since then, with more practical uses being developed worldwide. The study also identified challenges associated with GANs and suggested directions for future research.In addition to discussing the architecture and functionality of GANs, the paper highlighted their major applications, particularly in image segmentation, and the influence of deep learning in this area. It pointed out recent advancements in deep learning and the availability of large 3D CAD datasets, such as ShapeNet, which have driven significant progress in deep learning for object notation.The study provided a survey of GAN applications across various domains, giving an overview of how GANs are being used in different fields and emphasizing the technology's importance. The criteria for including papers in the review focused on application-based studies of GANs, progress in real-world industries, and alignment with the research questions and objectives of the study. The selection process included original research papers, reviews, and analytical studies published after 2016, with an emphasis on the latest advancements and applications of GAN models. (Aggarwal et al., 2021)

In medical imaging, using machine learning can be challenging because there aren't always enough labeled datasets. These datasets are expensive to create and require experts to carefully annotate each image. Annotating medical images is a slow process, which means there's not enough training data for supervised machine learning algorithms. This shortage of data can limit how well these algorithms perform. Generative Adversarial Networks (GANs) offer a solution to this problem. GANs can create synthetic (fake but realistic-looking) images that resemble real medical images. By generating these synthetic images, GANs help to expand the dataset available for training machine learning models. Researchers have shown that adding synthetic images created by GANs to the training data can significantly improve the accuracy of tasks like brain segmentation. In various studies, the Dice Similarity Coefficient (DSC), which measures how well segmented areas match with real data, has improved by 1 to 5 percentage points. This improvement is especially noticeable when there are only a few real training images available. It demonstrates that using GAN-created synthetic data alongside real data can boost the performance of machine learning models in medical imaging tasks.

There are many types of generative models, and one type called Generative Adversarial Networks (GANs) has become quite popular lately. However, evaluating and comparing GANs is not easy because there isn't a clear agreement on the best ways to measure how good they are.A recent paper looked at a wide range of methods for evaluating these models. It explored both quantitative (numbers-based) and qualitative (quality-based) measures to see which ones work best for assessing GANs. The goal was to find reliable ways to measure how well these models create new data, which is crucial for advancing generative modeling. The study suggested seven criteria to judge these evaluation methods. The idea was to figure out which approaches are most effective in giving accurate assessments of generative models. By carefully examining different ways to evaluate GANs, the paper aimed to tackle the challenges of comparing models created using GANs. Overall, the paper highlighted the importance of choosing the right evaluation methods. This helps ensure fair comparisons between different models and promotes progress in fields like computer vision and machine learning. (Borji, 2018)

BUILDING OUR FIRST GAN MODEL

Generative Adversarial Networks (GANs) have emerged as a powerful technique in deep learning, especially for creating data that closely resembles an existing dataset. This chapter delves into the core principles of GANs and provides a step-by-step guide to developing your first GAN model. By the conclusion of this chapter, you will have a clear understanding of GAN architecture, the training methodology, and how to implement a basic GAN model using Python and TensorFlow.

Setting Up the Environment

Before diving into building your first GAN model, you need to set up the appropriate environment:

1. **Install Python**: Verify that Python is installed on your system. Python is the primary programming language used for implementing GANs.
2. **Install Deep Learning Frameworks**: You will need either TensorFlow or PyTorch, which are popular deep learning frameworks. These libraries gives the necessary tools and functions to build and train neural networks.

3. **Create a Virtual Environment**: Setting up a virtual environment helps to manage dependencies and ensures that your project remains reproducible. You can create a virtual environment using virtualenv or conda.
4. **Install Required Libraries**: After activating your virtual environment, install TensorFlow or PyTorch along with other necessary libraries like NumPy and Matplotlib.

Defining the Generator

The Generator is a neural network designed to create synthetic data samples that closely resemble real data.

1. **Input Noise Vector**: The Generator begins with input of randomly generated noise. This noise is usually sampled from a standard normal distribution.
2. **Network Layers**: The Generator consists of several layers that progressively transform the input noise into a data sample. These layers can include dense (fully connected) layers, convolutional layers, and upsampling layers, depending on the type of data being generated.
3. **Activation Functions**: Activation functions such as ReLU (Rectified Linear Unit) and tanh (hyperbolic tangent) are applied within network layers to introduce non-linearity, enabling the model to capture intricate patterns and relationships in the data.

Defining the Discriminator

The Discriminator is another neural network that differentiate between real and fake data:

1. **Input Data Samples**: The Discriminator receives data samples as input, which can be either real data from the dataset or fake data generated by the Generator.
2. **Network Layers**: The Discriminator typically uses convolutional layers for image data or dense layers for other types of data. These layers extract key information from the input data to help in making a decision.
3. **Activation Functions**: LeakyReLU and sigmoid are common activation functions used in the Discriminator. LeakyReLU helps in avoiding the dying ReLU problem, while the sigmoid function outputs a probability score indicating the likelihood of the data being real.
4. **Output**: The output is a probability value ranging from 0 to 1, where a score near 1 suggests authentic data and a score near 0 indicates synthetic data.

Training the GAN

In the GAN training process, an adversarial loop is established where the Generator and Discriminator are trained in turns.

1. **Adversarial Training Loop**: The GAN training process is iterative and consists of two main steps:

   Step 1: Training the Discriminator:

   - Feed a batch of real data samples to the Discriminator.
   - Feed a batch of fake data samples (generated by the Generator) to the Discriminator.
   - Calculate the Discriminator's loss based on its ability to classify real and fake data correctly.
   - Update the Discriminator's weights to minimize classification errors.

   Step 2: Training the Generator:

   - Generate a batch of fake data samples using the Generator.
   - Feed these fake samples to the Discriminator.
   - Calculate the Generator's loss based on the Discriminator's ability to classify the fake data as real.
   - Update the Generator's weights to improve its ability to fool the Discriminator.
2. **Alternating Training**: The training process alternates between updating the Discriminator and the Generator. This adversarial feedback loop helps both networks continuously improve.

Monitoring and Evaluating the Model

1. **Loss Functions**: Monitor the loss functions of both the Generator and Discriminator during training. Convergence of the loss values indicates that the networks are learning effectively.
2. **Quality of Generated Samples**: Regularly evaluate the quality of the generated samples by visually inspecting them or using quantitative measures such as Inception Score or Fréchet Inception Distance.

3. **Adjust Hyperparameters**: Based on the performance metrics, tweak hyperparameters like the learning rate, batch size, and network design to enhance the model's performance.

Building your first GAN model involves setting up the environment, defining the Generator and Discriminator architectures, and training them using an adversarial loop. By understanding these core concepts and following the steps outlined in this chapter, you can create a basic GAN model capable of generating new data samples

BUILDING MODELS FOR IMAGE GENERATION: GAN VS. NON-GAN APPROACHES

Building a model to generate images, is a complex process. When tackling such a task, we can use Generative Adversarial Networks (GANs) or non-GAN approaches, such as Variational Autoencoders (VAEs) or simple convolutional neural networks (CNNs). Here, we will discuss the steps involved in building both types of models, addressing the overfitting problem, and evaluating the results. Additionally, we will highlight the advantages of using GANs over non-GAN models.

Non-GAN Models (Handwritten Digits Generator Without GAN)

Convolutional Neural Networks (CNNs) are particularly well-suited for tasks involving image recognition, including the identification of handwritten digits. Let's explore the architecture, training process, and evaluation of a CNN model tailored for digit recognition using a dataset like MNIST.

Building the Model

1. **Architecture**:

A CNN model typically consists of multiple layers designed to extract and learn features from images. Here's a common architecture for a CNN used in digit recognition:

Input Layer: Accepts the grayscale images of handwritten digits.
Convolutional Layers: These layers perform convolution operations on the input images to identify key features. By utilizing filters, they are able to recognize patterns like edges, corners, and textures.

Activation Layers: Usually ReLU (Rectified Linear Unit) layers, which introduce non-linearity into the model, allowing it to learn more complex patterns.
Pooling Layers: Perform down-sampling (usually max pooling) to reduce the spatial dimensions of the feature maps, which helps in reducing the computational load and controlling overfitting.
Fully Connected Layers: After passing through several convolutional and pooling layers, the output is flattened and subsequently fed into one or more fully connected layers to perform the classification task.
Output Layer: Uses a softmax activation function to classify the images into one of the 10 digit classes (0-9).

Figure 3. Non-GAN model flowchart

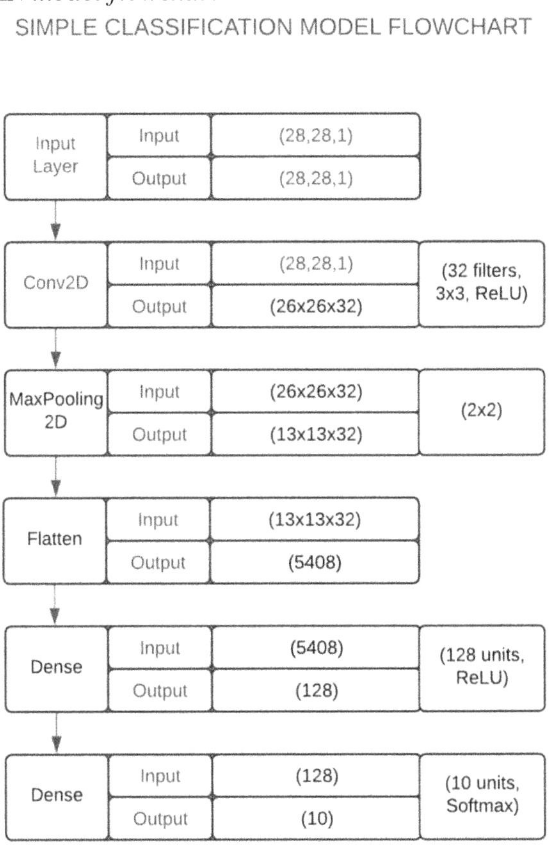

This (Figure 3) shows a flowchart for a simple convolutional neural network (CNN) used for image classification, like recognizing handwritten digits. Here's a brief explanation of each step:

Input Layer: Takes 28x28 grayscale images.
Conv2D Layer: Applies 32 filters (3x3) with ReLU activation, producing a 26x26x32 output.
MaxPooling2D Layer: Reduces the spatial dimensions by half, resulting in a 13x13x32 output.
Flatten Layer: Flattens the 13x13x32 output to a 5408-dimensional vector.
Dense Layer: Fully connected layer with 128 units and ReLU activation.
Dense Layer: Final output layer with 10 units (one for each class) and Softmax activation for classification.

2. **Training**:

Training a CNN involves the following steps:

Data Preparation: Preprocess the dataset by normalizing pixel values to a range of 0 to 1 and dividing it into training and validation sets.
Forward Propagation: Pass the input images through the network layers to get the predicted output.
Loss Calculation: Compute the loss (usually using categorical cross-entropy) by comparing the predicted output with the actual labels.
Backward Propagation: Calculate the compute the gradients of the loss concerning the model parameters through backpropagation.
Parameter Updates: Update the model parameters using an optimization algorithm like Adam or SGD to minimize the loss.
Iterate: Repeat the process for a number of epochs, passing through the entire training dataset multiple times to progressively improve the model's accuracy.

3. **Overfitting**:

Data Augmentation: Increase the diversity of the training set by applying random transformations (rotation, translation, etc.) to the images.
Dropout: Introduce dropout layers to randomly deactivate neurons during training, which prevents the model from becoming too reliant on specific paths.
Early Stopping: Monitor the validation loss and stop training when it starts to increase.
Regularization: Apply L2 regularization to the weights to penalize large weights, which helps prevent overfitting.

4. Evaluation

After training, the model's performance is evaluated using a separate test dataset. This involves:

Prediction: Running the test images through the trained CNN to get the predicted digit classes.
Accuracy Calculation: Comparing the predicted classes with the actual labels to compute the accuracy, which is the proportion of correctly classified images.
Confusion Matrix: Analyzing a confusion matrix to understand which digits are most commonly misclassified.
Additional Metrics: Optionally computing other metrics like precision, recall, and F1-score to get a more comprehensive view of the model's performance.

By following these steps, a CNN can be effectively trained and evaluated for the task of handwritten digit recognition, providing high accuracy and robustness against overfitting when properly tuned.

Results and Accuracy

This (Figure 4) graph illustrates how the overfitting problem is mitigated to some extent when using a CNN model, even without employing a GAN model.

Figure 4. Result of Non-GAN model

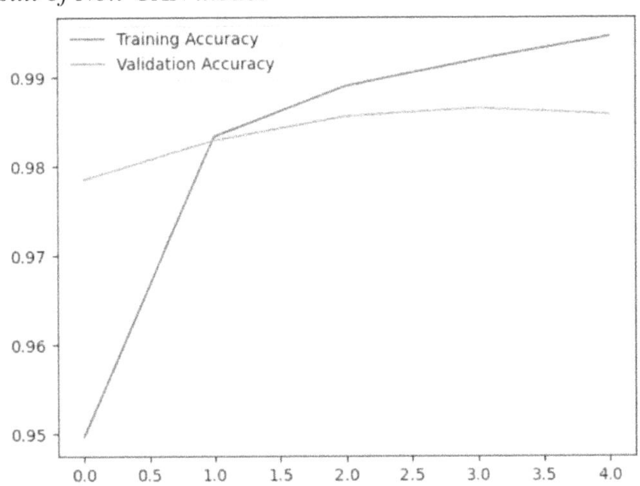

- **Quality**: VAE-generated images might be blurry compared to GANs.

- **Metrics**: Assessed using Inception Score (IS), Frechet Inception Distance (FID), and visual examination.
- **Overfitting**: Can occur if the model memorizes the training data and fails to generalize.

CNNs are highly effective for handwritten digit recognition due to their ability to automatically learn and extract relevant features from image data. By using a well-designed CNN architecture and proper training techniques, achieving high accuracy in identifying handwritten digits is feasible.

GAN Models (Handwritten Digits Generator With a GAN)

While GANs are primarily known for generating images, they can also be used to enhance image recognition tasks by learning the underlying data distributions. For recognizing handwritten digits, GANs can serve as auxiliary tools to boost the performance of a classification model. Here's an overview of how a GAN model can be adapted for recognizing handwritten digits from the MNIST dataset.

Building the Model

1. **Architecture**:

    Generator: Generates fake images from random noise. It uses fully connected layers followed by deconvolutional layers to create an image from a latent vector.

    Discriminator: Distinguishes between real and fake images. It uses convolutional layers followed by fully connected layers to classify the images.

Figure 5. GAN model flowchart

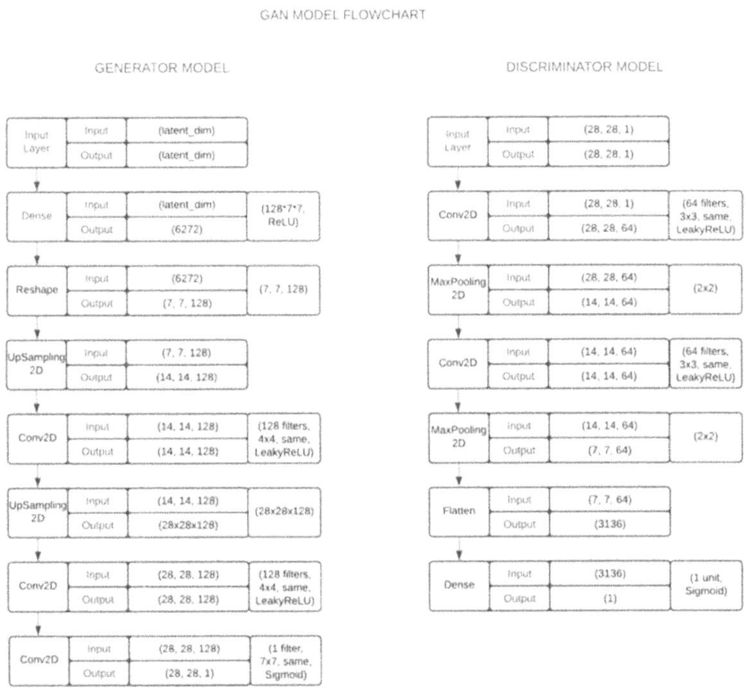

This image shows a flowchart for a Generative Adversarial Network (GAN) with two models: the Generator and the Discriminator. Here's a brief explanation:

Generator Model
Input Layer: Takes a latent vector of shape (latent_dim).
Dense Layer: Fully connected layer with 6272 units and ReLU activation.
Reshape Layer: Reshapes the output to (7,7,128).
UpSampling2D Layer: Upsamples to (14,14,128).
Conv2D Layer: Applies 128 filters (4x4) with LeakyReLU activation, maintaining shape (14,14,128).
UpSampling2D Layer: Up samples to (28,28,128).
Conv2D Layer: Applies 128 filters (4x4) with LeakyReLU activation, maintaining shape (28,28,128).
Conv2D Layer: Applies 1 filter (7x7) with Sigmoid activation to produce a final output image of shape (28,28,1).
Discriminator Model
Input Layer: Takes an image of shape (28,28,1).

Conv2D Layer: Applies 64 filters (3x3) with LeakyReLU activation, maintaining shape (28,28,64).
MaxPooling2D Layer: Reduces the dimensions to (14,14,64).
Conv2D Layer: Applies 64 filters (3x3) with LeakyReLU activation, maintaining shape (14,14,64).
MaxPooling2D Layer: Reduces the dimensions to (7,7,64).
Flatten Layer: Flattens the output to a vector of 3136 units.
Dense Layer: A dense layer with a single unit and a Sigmoid activation function, outputting a value that determines whether the input is real or fake.

2. **Training**:
   Data Preparation: Load and preprocess the MNIST dataset by normalizing the pixel values and preparing batches of data.
   Training the GAN:
   Discriminator Training: TTrain the Discriminator by using a combination of authentic and generated images. Calculate the loss using binary cross-entropy and adjust the Discriminator's parameters accordingly.
   Generator Training: Train the Generator to produce realistic images that can fool the Discriminator. This is done by updating the Generator's weights based on the Discriminator's feedback.
   Adversarial Training: Alternate between training the Discriminator and the Generator in each iteration to improve both components.
   Classification Model Enhancement:
   Data Augmentation: Use the trained Generator to create additional synthetic training data, which can augment the existing dataset.
   Feature Learning: Use the Discriminator's intermediate layers as feature extractors for a separate classification model. These features can improve the model's ability to recognize digits.
3. **Overfitting**:
   Data Augmentation: Apply random transformations to the training images to increase diversity.
   Label Smoothing: Smooth the labels for the discriminator to make it less confident and more robust.
   Noise Injection: Introduce noise to the discriminator's inputs to enhance its robustness.
   Batch Normalization: Incorporate batch normalization into both generator and discriminator to improve the stability of training and minimize the likelihood of overfitting.
4. **Evaluation**

Performance Assessment: Assess the effectiveness of the classification model by measuring metrics such as accuracy, precision, recall, and F1-score on a test dataset. using metrics like accuracy, precision, recall, and F1-score on a test set.
Comparison: Compare the performance of the augmented model (using GAN-generated data or features) with the baseline model to assess the improvements.

Using a dataset of handwritten digits, we can train a generative model to create new digit images. During training, an algorithm adjusts the model's parameters to reduce a loss function, allowing the model to learn the probability distribution of the training data. Once trained, the model can generate new digit samples, similar to those shown in the following (Figure 6):

Figure 6. Generative model being trained on handwritten digit samples and generating new digit images from random data inputs.

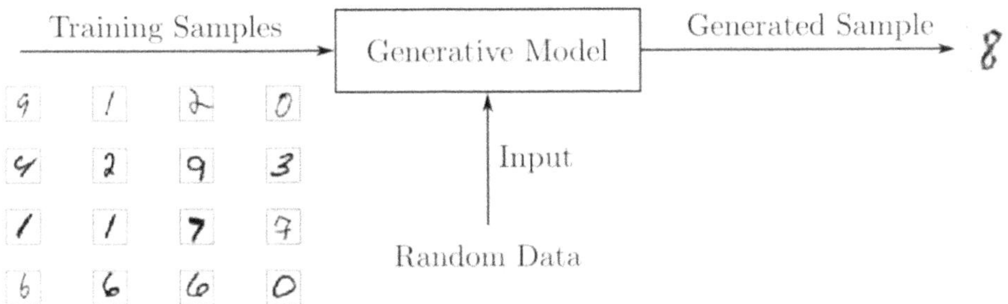

Here is a step-by-step implementation in Python using TensorFlow and Keras:

1. **Setup the environment:**

Make sure you have TensorFlow and Keras installed. If not, you can install them using pip.

Figure 7. Set up

```
pip install tensorflow
```

2. Import necessary libraries:

Figure 8. Import necessary libraries

```
import tensorflow as tf
from tensorflow.keras.layers import Dense, Reshape, Flatten, LeakyReLU, Dropout
from tensorflow.keras.models import Sequential
from tensorflow.keras.optimizers import Adam
import numpy as np
import matplotlib.pyplot as plt
```

3. Load and preprocess the MNIST dataset:

Figure 9. Load and preprocess

```
# Load the MNIST dataset
(X_train, y_train), (X_test, y_test) = mnist.load_data()

# Normalize the images to [0, 1]
X_train = X_train.astype('float32') / 255.0
X_test = X_test.astype('float32') / 255.0

# Reshape the images to (28, 28, 1) for the GAN
X_train = np.expand_dims(X_train, axis=-1)
X_test = np.expand_dims(X_test, axis=-1)

# Convert labels to categorical
y_train = to_categorical(y_train)
y_test = to_categorical(y_test)
```

4. Build the Generator:

Figure 10. Generator

```
# Build the generator
def build_generator(latent_dim):
    model = Sequential()
    model.add(Dense(128 * 7 * 7, activation='relu', input_dim=latent_dim))
    model.add(Reshape((7, 7, 128)))
    model.add(UpSampling2D())
    model.add(Conv2D(128, (4, 4), padding='same'))
    model.add(LeakyReLU(alpha=0.2))
    model.add(UpSampling2D())
    model.add(Conv2D(128, (4, 4), padding='same'))
    model.add(LeakyReLU(alpha=0.2))
    model.add(Conv2D(1, (7, 7), activation='sigmoid', padding='same'))
    return model
```

5, Build the Discriminator:

Figure 11. Build

```
# Build the discriminator
def build_discriminator():
    model = Sequential()
    model.add(Conv2D(64, (3, 3), padding='same', input_shape=(28, 28, 1)))
    model.add(LeakyReLU(alpha=0.2))
    model.add(MaxPooling2D((2, 2)))
    model.add(Conv2D(64, (3, 3), padding='same'))
    model.add(LeakyReLU(alpha=0.2))
    model.add(MaxPooling2D((2, 2)))
    model.add(Flatten())
    model.add(Dense(1, activation='sigmoid'))
    return model
```

5. Compile the GAN

Figure 12. GAN

```
# Compile the GAN
def compile_gan(generator, discriminator):
    discriminator.compile(optimizer=Adam(0.0002, 0.5), loss='binary_crossentropy', metrics=['accuracy'])
    discriminator.trainable = False
    gan_input = Input(shape=(latent_dim,))
    generated_image = generator(gan_input)
    gan_output = discriminator(generated_image)
    gan = Model(gan_input, gan_output)
    gan.compile(optimizer=Adam(0.0002, 0.5), loss='binary_crossentropy')
    return gan
```

6. Build and compile the GAN

Figure 13. GAN

```
# Build and compile the GAN
generator = build_generator(latent_dim)
discriminator = build_discriminator()
gan = compile_gan(generator, discriminator)
```

7. Train the GAN

Figure 14. Training

```
# Train the GAN
def train_gan(generator, discriminator, gan, epochs=1000, batch_size=64, interval=100):
    for epoch in range(epochs):
        # Train discriminator
        noise = np.random.normal(0, 1, (batch_size, latent_dim))
        generated_images = generator.predict(noise)
        real_images = X_train[np.random.randint(0, X_train.shape[0], batch_size)]
        labels_real = np.ones((batch_size, 1))
        labels_fake = np.zeros((batch_size, 1))
        d_loss_real = discriminator.train_on_batch(real_images, labels_real)
        d_loss_fake = discriminator.train_on_batch(generated_images, labels_fake)
        d_loss = 0.5 * np.add(d_loss_real, d_loss_fake)

        # Train generator
        noise = np.random.normal(0, 1, (batch_size, latent_dim))
        labels = np.ones((batch_size, 1))
        g_loss = gan.train_on_batch(noise, labels)

        # Print progress
        if epoch % interval == 0:
            print(f"{epoch} [D loss: {d_loss[0]} | D accuracy: {100 * d_loss[1]}] [G loss: {g_loss}]")

# Train the GAN
train_gan(generator, discriminator, gan)
```

The training process follows a repetitive adversarial loop. Below is a high-level pseudocode representation of training a GAN:

Figure 15. Each epoch

```
for each epoch do:
    Train Discriminator:
        - Sample real data batch
        - Generate fake data using Generator
        - Compute Discriminator loss
        - Update Discriminator weights

    Train Generator:
        - Generate fake data batch
        - Compute Generator loss (using Discriminator's feedback)
        - Update Generator weights
```

To prevent overfitting and ensure model stability, techniques like dropout, batch normalization, and feature matching can be used.

Training GANs requires substantial computing power. GPUs (e.g., NVIDIA RTX 3090) or cloud-based TPUs are recommended. Techniques such as gradient checkpointing and mixed-precision training can optimize memory usage and speed up training.

This code implements a Generative Adversarial Networks (GANs) implemented with TensorFlow and Keras to create images that mimic handwritten digits from the MNIST dataset. The process begins by setting up the environment with necessary imports and ensuring TensorFlow and Keras are installed. The MNIST dataset is loaded, normalized, and reshaped for compatibility with the GAN architecture. The Generator and Discriminator models are then defined: the Generator takes a latent vector as input and progressively up samples it through dense and convolutional layers to generate images, while the Discriminator employs convolutional layers followed by dense layers to differentiate between real and generated images. The GAN is compiled with separate optimizers for the Generator and Discriminator, where the Discriminator is initially trained on real and generated images, and subsequently, the Generator is trained to fool the Discriminator. Training occurs over multiple epochs, where batches of real and generated images are used to update the models, and progress is printed at regular intervals. Adjustments to parameters such as epochs and batch size can be made to refine the model's performance/

Table 2. Evaluation metrics for GANs

Metric	Purpose	Strengths	Weaknesses
Inception Score (IS)	Evaluates the quality of generated images	Measures image diversity	Does not measure realism accurately
Fréchet Inception Distance (FID)	Compares statistical similarity of generated and real images	More reliable than IS	Computationally expensive
Precision and Recall	Measures fidelity and diversity	Provides better insight into mode collapse	Harder to compute for large datasets

Evaluating GANs remains challenging due to their subjective nature. A hybrid approach combining multiple evaluation methods often yields the best results.

Results and Accuracy

The output should be numerical values that closely match the values in the training data, as demonstrated in Figure 7.

Figure 16. Output of handwritten digits generator with GAN

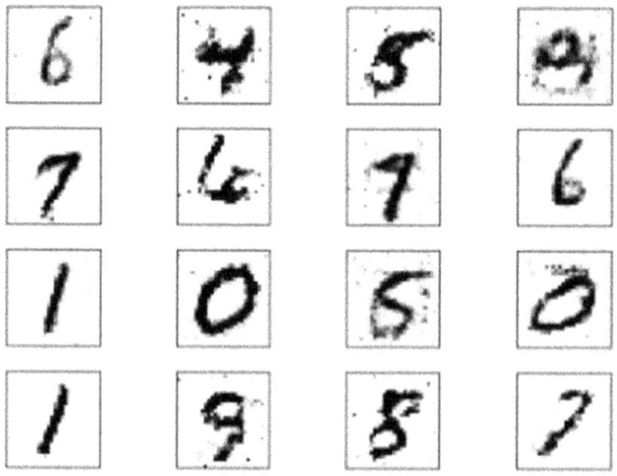

This (Figure 8) graph demonstrates how the overfitting problem can be mitigated using a GAN model. During the training phase, the GAN's two components, the Generator and the Discriminator, work against each other. The Generator creates new digit images, while the Discriminator evaluates their authenticity compared to real images. This adversarial process helps the model generalize better to new data, reducing overfitting and producing more realistic and varied digit samples.

Figure 17. Result of GAN model

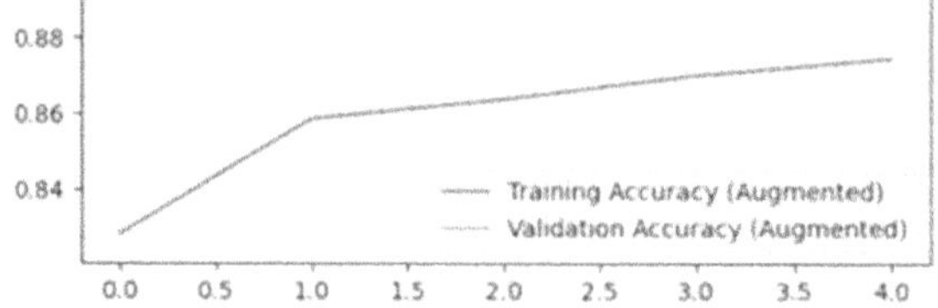

- **Quality**: GAN-generated images are often sharper and more realistic compared to VAE-generated images.
- **Metrics**: Evaluated using IS, FID, and visual inspection.

- **Overfitting**: Less prone to overfitting due to adversarial training but can still occur if the discriminator becomes too strong or the generator too weak.

ADVANTAGES OF USING GANS OVER NON-GAN MODELS

1. **Realism**: GANs produce higher-quality images that are more realistic and detailed compared to VAEs.
2. **Sharpness**: Images generated by GANs are typically sharper and less blurry than those generated by VAEs.
3. **Adversarial Training**: The dynamic between the generator and discriminator pushes the generator to improve continuously, leading to better outputs.
4. **Flexibility**: GANs can be adapted for different image generation tasks, such as image-to-image translation, super-resolution, and others.
5. **Diversity**: GANs often capture a wider variety of features in the data, leading to more diverse image generation.

Both GANs and non-GAN models like VAEs have their strengths and weaknesses in image generation tasks. GANs generally outperform VAEs in terms of image quality and realism, thanks to their adversarial training mechanism. However, managing the training stability and avoiding overfitting in GANs can be more challenging. Proper techniques and careful tuning are essential for both types of models to achieve the best results.

Applications Generative Adversarial Networks (GANs)

Generative Adversarial Networks (GANs) have numerous applications across various fields due to their capacity to produce authentic data. Here are several well-known applications of GANs:

1. **Image Synthesis and Generation:**

GANs are highly effective at creating new, realistic images by learning the distribution of a given dataset. This capability is used in generating high-resolution photographs, realistic avatars, and original artwork.

2. **Image-to-Image Translation:**

GANs can transform an image from one domain to another while preserving essential features. Examples include changing a daytime photo to nighttime, converting sketches into realistic images, and altering the artistic style of an image.

3. **Text-to-Image Synthesis:**

GANs can generate images based on textual descriptions. Given a text input, such as a phrase or caption, GANs can create corresponding visual content. This has significant implications for producing realistic images from text-based instructions.

4. **Data Augmentation:**

GANs can create synthetic data samples to augment existing datasets. This enhances the robustness and generalization of machine learning models by offering a broader range of training data.

5. **Data Generation for Training:**

GANs improve the detail and quality of low-resolution images by learning from sets of corresponding low-resolution and high-resolution images. This application is valuable in fields like medical imaging, satellite imaging, and video enhancement, where high-quality images are crucial.

6. **Video Synthesis:**

GANs can generate realistic videos or predict future frames in a video sequence, aiding in applications like video editing and predictive modeling.

7. **Art and Design:**

Artists and designers use GANs to generate creative content, providing inspiration and assisting in the creation of unique artistic pieces.

Table 3. GANs are widely used across industries

Application	Type of GAN Used	Outcome
Medical Imaging	SRGAN	Enhancing low-resolution medical scans
Climate Modeling	Conditional GAN	Generating climate predictions
Drug Discovery	GAN-based molecule generation	Discovering new pharmaceutical compounds
Video Game Development	StyleGAN	Creating realistic textures and assets

CONCLUSION AND FUTURE DIRECTIONS

Evaluating GAN models is essential for understanding how effectively they generate new data. This evaluation is crucial for advancing generative modeling in areas like computer vision and machine learning. The study underscores the importance of selecting the right evaluation methods to ensure fair comparisons between different models. By using appropriate evaluation criteria, we can promote progress in the field and improve the quality of generative models. The chapter explored various methods, both quantitative and qualitative, to determine the best ways to assess GANs. Understanding these evaluation techniques helps researchers accurately judge the performance of generative models. GAN variants like Conditional GANs (CGANs) and Laplacian Pyramid GANs (LAPGANs) offer unique capabilities for generating data based on specific needs. These advanced models build on the basic concepts of Vanilla GANs, demonstrating the evolution and flexibility of generative adversarial networks. The study also highlighted the practical applications of GANs in signal processing, showing the connection between theoretical knowledge and real-world use. This helps us effectively use GANs in their work by understanding both the theoretical framework and practical applications of these models.

The conclusion summarizes the chapter well but could better emphasize the future potential of GANs.

Emerging trends in GAN research include:

1. **GANs in Generative Design**: Used for creating innovative product designs.
2. **Integration with Reinforcement Learning**: Enhancing decision-making AI systems.
3. **Ethical Concerns**: Addressing deepfake misuse and bias in GAN-generated content.
4. **Scalability**: Improving training efficiency for large-scale applications.

REFERENCES

Aggarwal, A., Mittal, M., & Battineni, G. (2021). Generative adversarial network: An overview of theory and applications. *International Journal of Information Management Data Insights*, *1*(1), 100004. DOI: 10.1016/j.jjimei.2020.100004

Alqahtani, H., Kavakli-Thorne, M., & Kumar, G. (2019). Applications of generative adversarial networks (GANs): An updated review. *Archives of Computational Methods in Engineering*, *28*(28), 525–552.

Antipov, G., Baccouche, M., & Dugelay, J. L. (2017, September). Face aging with conditional generative adversarial networks. In 2017 IEEE international conference on image processing (ICIP) (pp. 2089-2093). IEEE.

. Bandi, A., Adapa, P. V. S. R., & Kuchi, Y. E. V. P. K. (2023). The power of generative AI: A review of requirements, models, input–output formats, evaluation metrics, and challenges.

. Bau, D., Zhu, J.-Y., Strobelt, H., Zhou, B., Tenenbaum, J. B., Freeman, W. T., & Torralba, A. (2021). GAN dissection: Visualizing and understanding generative adversarial networks.

Borji, A. (2018). Pros and cons of GAN evaluation measures. *Computer Vision and Image Understanding*.

Cheng, Z., & Li, W. (2022). Building more efficient GAN models: A comparative study. *Neural Networks Journal*, *64*, 210–220.

Creswell, A.Jr, White, T., Dumoulin, V., Arulkumaran, K., Sengupta, B., & Bharath, A. A. (2018). Deep learning for visual understanding: Part 2. *IEEE Signal Processing Magazine*, •••, 53–55. DOI: 10.1109/MSP.2017.2765202

Dutta, I. K., Ghosh, B., Carlson, A., Totaro, M., & Bayoumi, M. (2020, October). Generative adversarial networks in security: A survey. In 2020 11th IEEE Annual Ubiquitous Computing, Electronics & Mobile Communication Conference (UEMCON) (pp. 0399-0405). IEEE.

Goodfellow, I., Pouget-Abadie, J., Mirza, M., Xu, B., Warde-Farley, D., Ozair, S., & Bengio, Y. (2020). Generative adversarial networks. *Communications of the ACM*, *63*(11), 139–144.

Gupta, R., & Sharma, A. (2021). GANs for 3D object generation and simulation: A review. *Journal of Computational Intelligence*, *33*(6), 876–889.

Hao, Y., Wang, X., & Li, Q. (2021). Variants of GANs for data augmentation: An overview. *Journal of Machine Learning Research, 45*(3), 212–230.

Kim, D., & Park, C. (2021). Applications of GANs in creative industries: Art generation and design. *Computers and Creativity, 12*(5), 77–89.

Li, Y., & Chen, M. (2023). Investigating the convergence in GANs: Mathematical and empirical insights. *Journal of Neural Engineering*, 19(2), 055-070.

Lin, Y., Wei, Z., & Liu, H. (2023). A study on the effect of training data on the performance of GAN models. *Machine Learning Research, 48*(2), 32–50.

Liu, S., Xu, Y., & Zhao, H. (2022). A comprehensive review on the application of GANs in video generation. *International Journal of Computer Vision and Graphics, 29*(3), 183–201.

Liu, W., Zhang, J., & Wei, Y. (2023). A deep dive into the generator architecture of GANs for realistic image generation. *IEEE Transactions on Image Processing, 35*(6), 321–335.

Nayak, A. A., Venugopala, P. S., & Ashwini, B. (2024). A systematic review on generative adversarial network (GAN): Challenges and future directions. *Archives of Computational Methods in Engineering*, April.

Patel, A., & Kaur, S. (2022). Applications of GANs in healthcare. *IEEE Transactions on AI, 12*(4), 214–230.

Saxena, D., & Cao, J. (2021). Generative adversarial networks (GANs) challenges, solutions, and future directions. *ACM Computing Surveys, 54*(3), 1–42.

Singh, P., & Gupta, S. (2022). Unsupervised learning in GANs: The quest for high-quality image synthesis. *IEEE Transactions on Neural Networks and Learning Systems, 33*(5), 788–802.

Smith, J., & Lee, R. (2023). Advances in GAN training. *Journal of AI Research, 58*(3), 145–162.

. Suryadevara, C. K., & Wilmington University. (2020). Generating free images with OpenAI's generative models. *International Journal of Virtual Technology and Multimedia*, 49-50.

Vaidya, R., Kumar, P., & Singh, G. (2022). Survey on conditional GANs for image-to-image translation. *Artificial Intelligence Review, 44*(2), 231–246.

. Wang, C., Xu, C., Yao, X., & IEEE. (2019). Evolutionary generative adversarial networks. *Journal Name*, Volume(Issue), pages.

. Wang, Z., She, Q., & Ward, T. E. (2021). Generative adversarial networks in computer vision: A survey and taxonomy.

Xu, J., & Liu, Y. (2023). Evolution of GAN architectures for facial recognition and synthesis. *Computer Vision and Image Understanding*, *207*, 102215.

Yang, Q., & Zhang, T. (2022). From GANs to adversarial learning: A new perspective on optimization. *Journal of Optimization Theory and Applications*, *59*(1), 105–120.

Yao, X., Zhang, M., & Wang, Y. (2022). Real-time generation of deepfake content using GANs: Challenges and ethical implications. *Journal of AI Ethics*, *9*(4), 45–60.

. Yi, X., Walia, E., & Babyn, P. (2019). Generative adversarial network in medical imaging: A review. *Journal Name*, Volume(Issue), pages.

Zhang, K., Zhao, X., & Li, H. (2022). Exploring deep learning models for image generation using GANs. *Journal of Artificial Intelligence and Image Processing*, *7*(4), 56–67.

Zhang, L., & Shen, H. (2022). Image-to-image translation with conditional GANs: An investigation. *International Journal of Computer Vision*, *47*(2), 123–139.

Zhang, L., & Zhang, H. (2023). Enhancements in generative adversarial networks for fashion design and clothing generation. *Fashion Technology Journal*, *5*(1), 42–58.

Zhou, X., & Chen, Y. (2021). Hybrid generative models for image synthesis. *Neural Computation Journal*, *29*(2), 99–121.

Zhu, M. (2021). A brief analysis of GAN variants on image classification and generation. *Journal of Physics: Conference Series*, *1827*(1), 012165. DOI: 10.1088/1742-6596/1827/1/012165

Chapter 9
Improving GANs for Text-to-Image With Secondary Loss Functions and Enhanced Attention Mechanisms

Deepika Roselind Johnson
https://orcid.org/0000-0002-3610-2979
Vellore Institute of Technology, Chennai, India

Logeswari Govindaraj
Vellore Institute of Technology, Chennai, India

ABSTRACT

Generative Adversarial Networks (GANs) have emerged as versatile tools for synthesizing content across multiple domains. Despite their capability to produce realistic outputs, ongoing research endeavors to further enhance their quality. Key objectives in GAN development encompass reducing training loss and enhancing performance metrics like Inception Score and Fréchet Inception Distance (FID). This paper introduces a novel method to enhance the quality of generated images from textual descriptions by incorporating a secondary loss function. The proposed approach focuses on refining the alignment between images and text during training, aiming to improve the fidelity of synthesized outputs. Leveraging the KT-GAN framework, which extends AttnGAN by integrating attention mechanisms for relevance computation, our method seeks to fine-tune the image-text matching process. Detailed descriptions of the existing KT-GAN system and the proposed enhancements are provided in subsequent sections, demonstrating the potential for

DOI: 10.4018/979-8-3693-7152-7.ch009

significant advancements in GAN-based image synthesis.

I. LITERATURE REVIEW

Generating high-resolution images from text captions presents a formidable challenge, yet it holds immense significance across various practical domains such as art generation and computer-aided design. Recent advancements in deep generative models have significantly propelled progress in this domain. Mansimov's alignDRAW model (Mansimov et al., 2016) represents an extension of the Deep Recurrent Attention Writer (DRAW) (Gregor et al., 2015), enabling iterative drawing of image patches while dynamically attending to pertinent words in the caption. Additionally, the authors (Nguyen et al., 2017) proposed an approximate Langevin approach for generating images from captions. In contrast, authors (Reed et al., 2017) employed a conditional PixelCNN (Oord et al., 2016) with a multi-scale model structure to synthesize images from text.

In contrast to alternative deep generative adversarial models, Generative Adversarial Networks (GANs) (Gregor et al., 2015) have exhibited remarkable efficacy in producing sharper samples. Authors (Reed et al., 2016) were pioneers in illustrating the capacity of conditional GANs to generate credible images from textual descriptions. Subsequent studies by the same authors showcased further advancements, highlighting GANs' capability to enhance sample quality through the inclusion of additional conditions, such as object locations. Authors (Zhang et al., 2017) implemented several stacked GAN architectures for text-to-image synthesis, employing diverse GAN models to generate images of varying sizes.

While previous GAN architectures have primarily relied on conditioning via global sentence vectors, they have lacked the granularity provided by word-level information necessary for precise image generation. In recent years, attention mechanisms have emerged as essential components of sequence transduction models, demonstrating success in capturing multi-level dependencies across various tasks such as image captioning, question answering, and machine translation. Researchers (Vaswani et al., 2017) showcased the effectiveness of attention models in achieving high-quality results in machine translation tasks. Despite these advancements, the integration of attention mechanisms into GANs for text-to-image synthesis remains unexplored. Notably, while alignDRAW (Masimov et al., 2016) employed LAPGAN (Denton et al., 2015) to enhance image resolution, their GAN framework was utilized solely as a post-processing step, devoid of attention mechanisms. Introducing the AttnGAN represents a groundbreaking development, as it pioneers the integration of attention mechanisms into GANs, enabling the generation of fine-grained, high-quality images through multi-level conditioning encompassing both word and sentence levels.

Attention mechanisms have found widespread application in both computer vision and natural language processing domains, notably in tasks such as object detection, image/video captioning, person re-identification, and visual question answering. In the realm of Text-to-Image (T2I) synthesis, recent advancements include the introduction of word-level visual attention mechanisms by AttnGAN (Tao et al., 2020). This innovation enhances the synthesis process by focusing on fine-grained details across various regions of the generated image. Subsequent work, exemplified by obj-GAN (Li et al., 2019), introduces an object-driven attention mechanism aimed at further refining detail synthesis, thereby producing even more intricate and refined images.

Both AttnGAN (Xu et al., 2019) and obj-GAN (Li et al., 2019) exhibit a limitation in their fixed word weights, which can lead to the neglect of important words by the attention mechanisms. Recently, SEGAN (Hongchen et al., 2019) introduced an attention regularization term to address this issue by filtering out unimportant words and emphasizing significant ones. However, determining the adaptive threshold in the regularization term poses challenges. Another recent approach (Zhen et al., 2019) proposed an attention update mechanism for transferring the pose of a person, which could leverage appearance features to guide the pose transfer process seamlessly.

Inspired by (Zhen et al., 2019), a novel attention update mechanism aimed at dynamically focusing on crucial words is proposed. This mechanism, termed as the word feature update module, calculates word features using a normalized cross-correlation matrix, establishing correlation between image and word features from the preceding layer. These word features are then updated using a method outlined by (Tan et al., 2020), involving an element-wise product of the reshaped normalized cross-correlation matrix and the word features from the previous layer. This approach yields a more precise encoded representation of the text captions, thereby improving the quality of the generated visual representations.

III. PROPOSED WORK

The proposed approach introduces a system architecture comprising four main modules, each with sub-modules. In Figure 1, the pre-existing components from the previously proposed KT-GAN (Tan et la., 2020) are specified in green, while the additional enhancements from the proposed approach are highlighted in pink.

Figure 1. Architecture of the proposed system

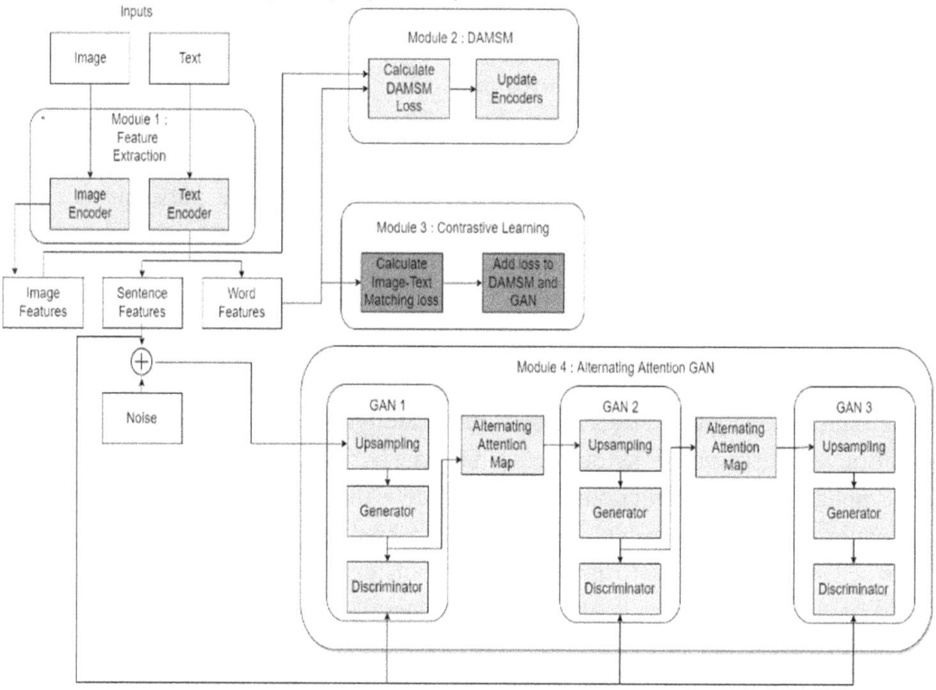

The system architecture begins with Module 1 shown in Figure 1, comprising two models: an image encoder and a text encoder, implemented as a Bi-LSTM and CNN respectively. These encoders generate word, sentence, and image features, which undergo fine-tuning processes in Module 2 and Module 3, as depicted in Figure 1. Additionally, a noise vector is incorporated into the sentence features, which are then linked to Module 4 in Figure 1 to introduce variability in the generated samples. The core of the system architecture lies within Module 4 in Figure 1, housing the principal components of the GAN: the generator and discriminator. Text and image encoders produce text features (word, sentence) and image features, respectively. Following this, a noise function is applied to these features to introduce variation during both generation and training. This variability is essential for the discriminator to effectively differentiate between fake and real samples.

The DAMSM module (Deep Attentional Multimodal Similarity Module), featured in Module 2 of Figure 1, serves as a crucial loss function applied during encoder training, aimed at extracting precise features from the dataset. Contrastingly, the highlighted enhancement from the proposed approach, the contrastive learning module, found in Module 3 of Figure 1, is designed to compute a similarity score between images, effectively grouping similar samples while distinguishing dissimilar ones.

Contrastive learning, as introduced in Module 3 of Figure 1, addresses this challenge by training the model to discern between similar and dissimilar data points, thereby learning general features of the dataset without relying on labels. This principle underpins the current study's objective of enhancing the performance of the existing KT-GAN (Tan et al., 2020) model by integrating contrastive learning techniques. The initial approach to attention-driven generative networks, as seen in AttnGAN (Xu et al., 2019), introduced a set of fundamental modules, which have remained foundational in subsequent approaches. Enhancements in KT-GAN (Tan et al., 2020) have introduced additional modules while retaining the core modules from the original design. Building upon this, the proposed approach in this study maintains the same core modules while introducing improvements in an additional module.

Refined Feature Extraction Via Bi-LSTM and CNN

Text features are derived from the input text via a Bi-LSTM (RNN) text encoder, while image features are extracted from input images using a CNN image encoder. These extracted features are utilized in training the GAN model and computing the DAMSM loss, as well as the contrastive loss from the contrastive learning module. Within the Bi-LSTM, each word is linked with two hidden states, which are subsequently combined to represent the word's meaning. Meanwhile, the image encoder generates image features from images, with certain layers focusing on local features and others on global features. This module serves as a fundamental component in both the original AttnGAN (Xu et al., 2019) and subsequent approaches, as feature extraction is indispensable in the training of any generative networks.

Figure 2. Feature extraction module

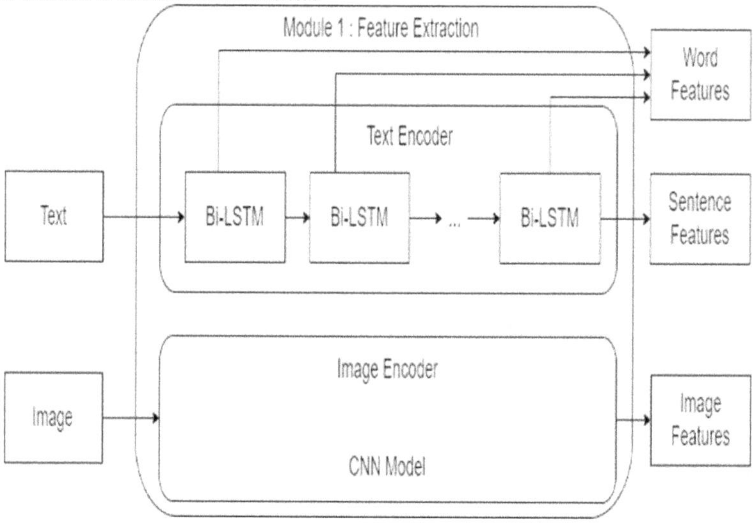

As illustrated in Figure 2, the Bi-LSTM constructs networks with loops, enabling the persistence of information over sequential input. This looped structure enables the neural network to process input sequences effectively. Additionally, it transforms independent activations into dependent activations by sharing the same weights and biases across all layers. Consequently, this reduces parameter complexity and avoids the need to memorize each previous output, as each output serves as input to the subsequent hidden layer. Therefore, these three layers can be amalgamated into a single recurrent layer, where the weights and biases remain consistent across all hidden layers.

$$h_t = f\left(h_{t-1}, X_t\right) \tag{1}$$

Equation (1) computes the current state of the Bi-LSTM network. The variable h_t denotes the current state.

$$h_t = \tanh\left(W_{hh} h_{t-1} + W_{xh} X_t\right) \tag{2}$$

The subsequent stage involves applying the tanh activation function. Equation (2) computes the current state after the activation function is applied. Here, h_t represents the current state, W_{hh} denotes the weight at the recurrent neuron, and W_{xh} signifies the weight at the input neuron.

$$y_t = W_{hy} h_t \tag{3}$$

Equation (3) computes the output of the Bi-LSTM network, represented by the variable y_t. The weight of the output layer is denoted by W_{hy}. However, in a Bi-LSTM, where both backward and forward activation functions are incorporated, the expression is modified accordingly, as depicted in Equation (3).

$$y_t = g\left(W_y \left[a^{ft} + a^{bt}\right] + b_y\right) \quad (4)$$

In Equation (4), which incorporates both backward and forward activations, a hyperparameter g influences the strength of the output. The product of the sum of forward and backward activations, represented as $[a^{ft}+a^{bt}]$ and the weight of the output layer W_y.

In the conventional convolutional neural network, data flow is delineated through both forward and backward propagations. These propagations consist of convolutional and max-pooling layers. Mathematically, the convolutional layer in forward propagation can be expressed as,

$$x_{ij}^{l} = \sum_{a=0}^{m-1}\sum_{b=0}^{m-1} \omega_{ab} y_{(i+a)(j+b)}^{l-1} \quad (5)$$

Max-pooling layers are straightforward and do not undergo learning themselves. Their function is to take a k×k region and output a single value, which corresponds to the maximum value within that region. For example, if the input layer is of size N×N, the max-pooling layer will output a layer of size N/k × N/k, as each k×k block is condensed to a single value using the max function.

During forward propagation, k×k blocks are condensed to a single value. Subsequently, this single value is assigned an error computed from backward propagation from the preceding layer. This error is then transmitted back to its origin. Since it originated from only one location within the k×k block, the backpropagated errors from max-pooling layers tend to be sparse.

Deep Attentional Multimodal Similarity Model (DAMSM)

The objective of the DAMSM model is to compute an image-text matching score, indicating the similarity between the input text and the generated image output. The DAMSM loss function calculates a loss value based on the disparity between the generated image and the input sentence. This loss is then utilized to train the encoders, ensuring that they produce features that more accurately capture the semantic correspondence between the two modalities. The goal is to minimize the DAMSM loss, thereby maximizing the similarity scores between the images and their associated descriptive input sentences.

The DAMSM loss is structured to train the attention model in a semi-supervised manner, where the sole supervision is provided by the correspondence between entire images and complete sentences (sequences of words). As depicted in Figure 3, for a batch of image-sentence pairs $\{(Q_i, D_i)\}_{i=1}^{M}$, the posterior probability of sentence D_i matching with image Q_i is calculated as follows.

$$P\left(Di \middle| Qi\right) = \frac{exp(y_3 3R(Q_i, D_i))}{\sum_{j=1}^{M} exp(y_3 R(Q_i, D_j))} \tag{6}$$

Figure 3. Deep attentional multimodal similarity model

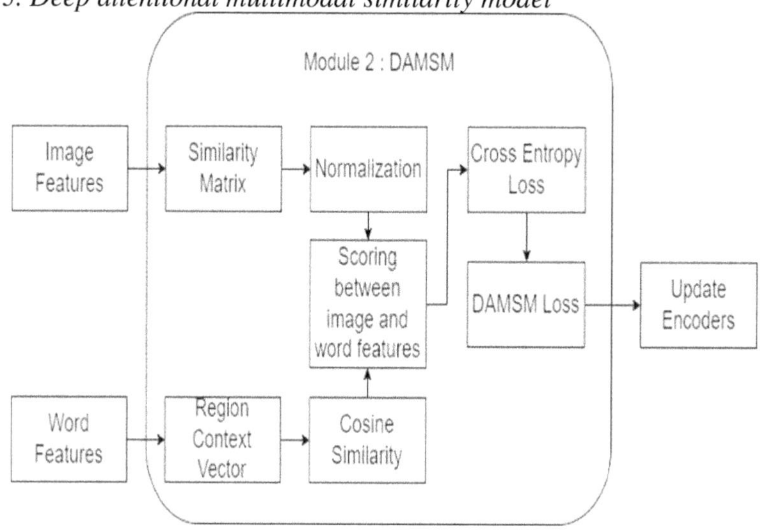

Equation (6) computes the posterior probability that sentence D_i is associated with its corresponding image Q_i. The loss function is characterized by the negative logarithm of the posterior probability, representing the likelihood that the images are aligned with their respective text descriptions, expressed as,

$$L_1^w = -\sum_{i=1}^{M} logP\left(D_i \middle| Q_i\right) \tag{7}$$

Similarly, in a symmetric manner, the function can be minimized by,

$$L_2^w = -\sum_{i=1}^{M} logP\left(Q_i \middle| D_i\right) \tag{8}$$

Here, the logarithm of the posterior probability from Equation (6) is computed to derive the second loss. By substituting the attention-driven text-matching scores from Equations (7) and (8), secondary losses L_1^s and L_2^w are derived. By summing up all these subsequent losses, the overall DAMSM loss is formulated.

$$L_{DAMSM} = L_1^{\,w} + L_2^{\,w} + L_1^{\,s} + L_2^{\,s} \tag{9}$$

Contrastive Learning

The primary concept behind contrastive learning is to simultaneously train representations that exhibit similarity. Introducing contrastive loss during image-text matching enhances the coherence among generated images corresponding to the same image description. The encoders are expanded to generate image-text matching loss, specifically contrastive loss. Contrastive learning typically encompasses three components: data sampling, the core model architecture, and the introduction of contrastive loss within the loss function. During each training epoch, a minibatch of images is sampled, from which the text and image encoders extract features that are then matched to each other. The contrastive loss function aims to minimize the distance between similar fake images while maximizing the distance between different fake images.

Figure 4. Contrastive learning model

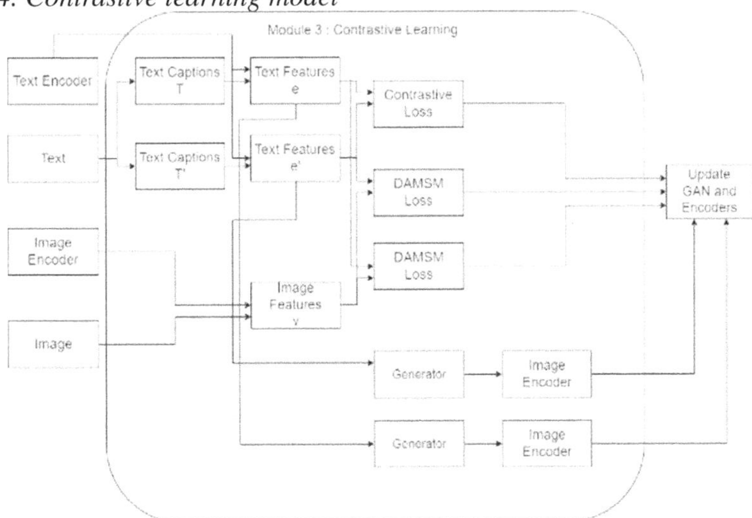

In Figure 4, the DAMSM loss function serves as the baseline for generating an image-text matching loss. To generate the contrastive loss, the Normalized Temperature-scaled Cross Entropy Loss (NT-Xent) is selected. For a given pair, where sim(a, b) = $a^T b/\|a\|\|b\|$ denotes the dot product between l2 normalized vectors a and b, the loss function for the ith sample is defined as follows.

$$L\left(i\right) = -\frac{\log(\exp(sim(ui,uj)/\tau)}{\sum_{k=1}^{2N}\log\exp(sim(ui,uk)/\tau)} \quad (10)$$

The overall contrastive loss is calculated across all positive pairs in a minibatch, and it can be expressed as.

$$L_c = \frac{1}{2N}\sum_{i=1}^{2N} L\left(i\right) \quad (11)$$

Alternating Attention GAN

The fundamental GAN architecture comprises a generator and a discriminator as shown in Figure 5. Features extracted using text and image encoders trained in previous modules are employed in training the GAN and generating attention maps, forming the foundation of the AttnGAN. Initially, the latent space consists of noisy values ranging from 0 to 1, indicated by the noise vector. As it traverses through the layers of the generative network, the latent space undergoes updates, enhancing the quality of the generated image with the aid of attention maps. Attention map visualizations depict the extent of each text feature's influence, while highlighting the strength of the corresponding text features. However, in the base AttnGAN model, the word features used to construct attention maps remain unimproved. This underscores the rationale behind introducing a mechanism aimed at enhancing the word features themselves, thus constituting an improved version of the base AttnGAN model.

In Figure 5, the word feature update module comprises three primary steps. The first step involves computing an NCC matrix R* between the image features H_i and word features W_i. The word features are projected into the same latent semantic space as the image features using a perceptual layer. Each column of the hidden features represents a feature vector of an image's sub-region. Subsequently, the next step is to calculate a cross-correlation matrix $R_{i,j}$, which captures the dot-product similarity between the ith word in the sentence and the jth sub-region in the image. Thus, the computed NCC matrix can be described as.

$$R^* = {r^*}_{i,j} = \frac{exp\left(r_{i,j}\right)}{\sum_{k=1}^{T} exp\left(r_{k,j}\right)} \quad (12)$$

Figure 5. Alternating attention GAN module

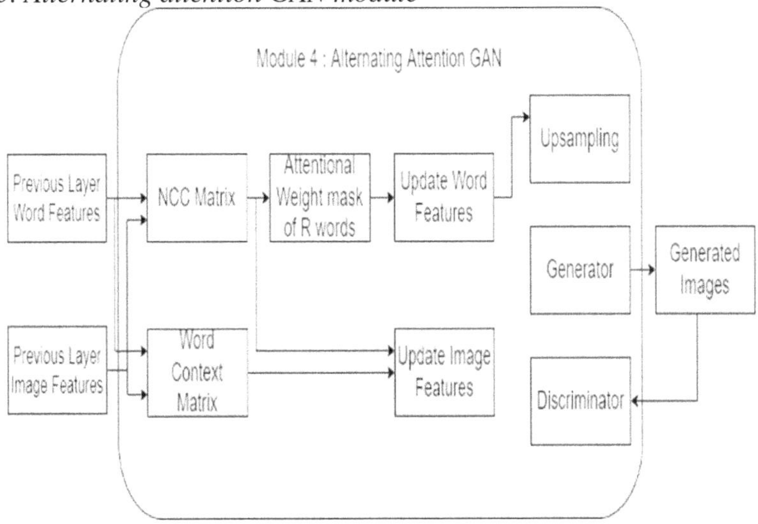

The NCC matrix R* is subsequently reshaped through max-pooling to yield the attentional weight mask of words, denoted as R^, by duplicating each column the same number of times as present in the mask. Finally, the word feature is updated as follows.

$$W_i = \alpha \cdot [R^{\wedge}. W_i - 1] \oplus \beta \cdot [W_i - 1], (i = 1, 2, ...m - 1) \tag{13}$$

Equations (12) and (13) determine the NCC matrix and update the word features, respectively. This process is iterated for each forward pass of the GAN neural network throughout the training phase.

RESULTS

The dataset employed in this study is the Caltech UCSD Birds 200 (CUB-200) dataset, which contains images depicting various bird species alongside corresponding captions. It encompasses a collection of photographs showcasing 200 distinct bird species, predominantly those indigenous to North America. Within this dataset, annotations are provided for each image, including details such as bounding boxes, rough segmentation, and attributes. Additionally, each image is associated with ten captions, further enriching the dataset with textual descriptions. Furthermore, the dataset includes annotations for fifteen-part locations and 312 binary attributes, offering comprehensive information about the depicted birds.

Figure 6. Sample training images from the dataset

Figure 6 showcases a selection of sample images from the dataset utilized for training, each accompanied by detailed annotations including 1 subcategory label, 15 part locations, 312 binary attributes, and 1 bounding box. In augmenting the textual information within the CUB-200-2011 dataset, fine-grained natural language descriptions have been collected, comprising ten single-sentence captions for each image. The Bi-LSTM architecture incorporates bidirectional input flow, distinguishing it from standard LSTM models. While traditional LSTMs enable input flow in either a backward or forward direction, bidirectional LSTMs allow for input flow in both directions, facilitating the retention of both past and future information. This capability proves beneficial for tasks such as predicting missing elements in sentences based on contextual information.

The image encoder, derived from a CNN model, comprises convolutional layers, max-pooling layers, and mixed layers sourced from the Inception V3 model. Before entering the CNN, the input image undergoes reshaping, with the CNN subsequently extracting features within the embedding dimension. These extracted word features, illustrated in Figure 6 and Figure 7, are shaped to align with the image features extracted from the CNN, facilitating feature correlation. The initial item in both arrays signifies the number of input images processed by the image encoder (3), resulting in three corresponding output matrices. The second item in the output tensor (256) denotes the embedding dimension, consistent with both encoders, typically set to

300 or a power of 2. Reshaping the input image shape to 64x64 reduces training time, while the 17x17 output shape represents the dimensions of the output features post-convolutional, max-pooling, and mixed layers of the CNN.

The Deep Attentional Similarity Model (DAMSM) functions as a loss function, aiding in encoder fine-tuning by aggregating word and sentence losses derived from encoder-extracted features. The DAMSM loss, comprising four losses is recorded at the 100th training epoch. Typically, as training progresses, loss values decrease, indicative of model improvement. Also, the application of contrastive learning loss to the DAMSM loss is calculated from word and text features, further refining features. This loss stems from the correlation matrix of sentence embeddings obtained through an L2-Norm operation on original sentence features extracted during training. Upon calculating the correlation matrix between S1 and S2 and reshaping based on the attentional weight mask, the contrastive loss is derived.

In the Alternating AttnGAN model, the core GAN architecture comprising a generator and discriminator remains consistent. These components leverage fine-tuned features from preceding layers to predict images based on provided text inputs. For instance, sample inputs such as "the bird is completely red with black wings and a pointy beak," "the bird has a blue crown with white throat and brown secondaries," and "a yellow bird with grey back, yellow tarsus and brown throat, nape with a black face" are provided to the GAN model. The corresponding outputs from GANs trained over various epochs along with their corresponding attentions maps are illustrated in Figure 7 and Figure 8, these outputs demonstrate the evolution and improvement of image generation capabilities across different training stages of the GAN model.

To assess the effectiveness of generative models, performance metrics play a pivotal role. Among these metrics, two key indicators stand out: the Inception Score (IS) and the Fréchet Inception Distance (FID). These metrics provide valuable insights into the quality and diversity of generated samples. Additionally, monitoring training losses and learning rates over the course of GAN and encoder training epochs serves as an important supplementary metric, offering deeper understanding and evaluation of model performance.

Figure 7. The first row indicates output of GAN from 10 epochs of training and the second row indicates output of GAN from 100 epochs of training

Figure 8. The respective attention maps for the inputs

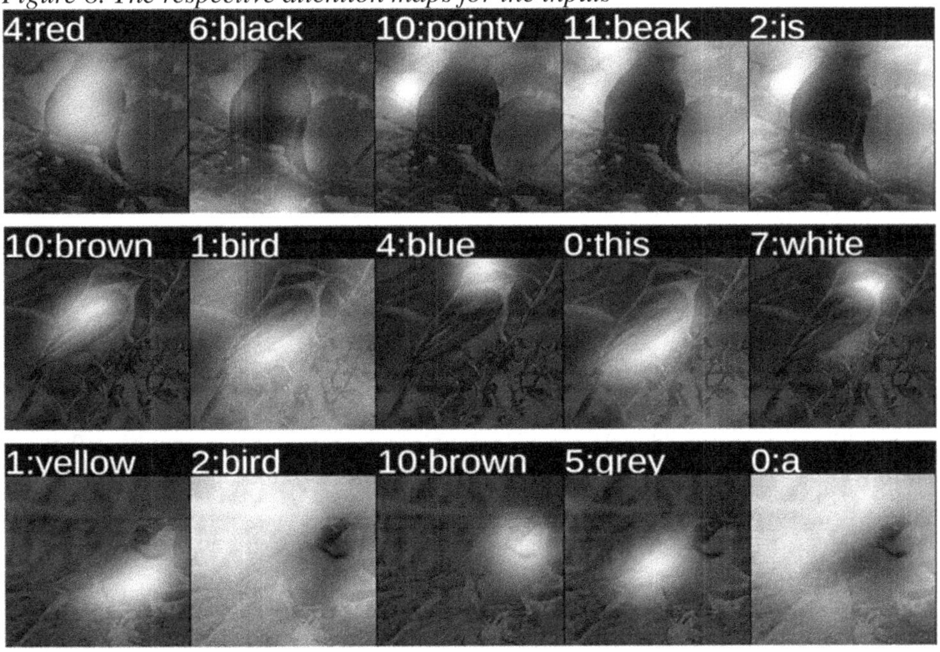

Inception Score

The Inception Score evaluates both the quality and diversity of generated images. It measures the predictability of object types in synthetic images, with high predictability indicating good image quality. This score is derived using an inception network to classify synthetic images and predict their labels. Specifically, the score is calculated as the marginal probability of the predicted labels given the synthetic images, expressed as P(y|x), where y represents the label and x denotes the synthetic image.

$$\int (y|x = G(z))\, dz$$

The diversity of generated images is assessed by examining the uniformity of the data distribution for the predicted labels, ensuring high entropy. This diversity aspect is incorporated into the Inception Score calculation by measuring the Kullback-Leibler divergence.

$$IS\left(G\right) = exp\left(E_{x\sim pa} D_{KL}\left(p\left(y|x\right) \| p\left(y\right)\right)\right) \tag{14}$$

Fretchet Inception Distance (FID)

The Fretchet Inception Distance (FID) evaluates the dissimilarity between the feature representations of real and generated images. It measures the squared difference between the mean feature vectors, denoted as $||m\text{-}m_w||^2$, and incorporates the covariance matrices of the feature vectors, represented as C and C_w. By calculating the trace of the matrix expression $(C+C_{w\text{-}2}{}^* (C^*C_x))$, where C_x represents the covariance matrix of the joint distribution, and adding it to the squared difference, the FID score is computed.

$$d^2((m, C), (m_w, C_w)) = \|m - m_w\|_2^2 + Tr(C + C_w - 2(CC_w)^{1/2}) \tag{15}$$

The model's performance is assessed using the Inception Score and FID metrics as shown in Table 1, where 500 generated images are sampled from random input sentences within the dataset's context.

Table 1. Performance analysis for trained GAN model

Data Used	Inception Score	Fretchet Inception Distance
Validation set - 500 generated images	Mean = 4.71	-
Dataset - 11,788 images Validation Set - 500 generated images	-	17.58

Figure 9. Tracking word loss versus epochs

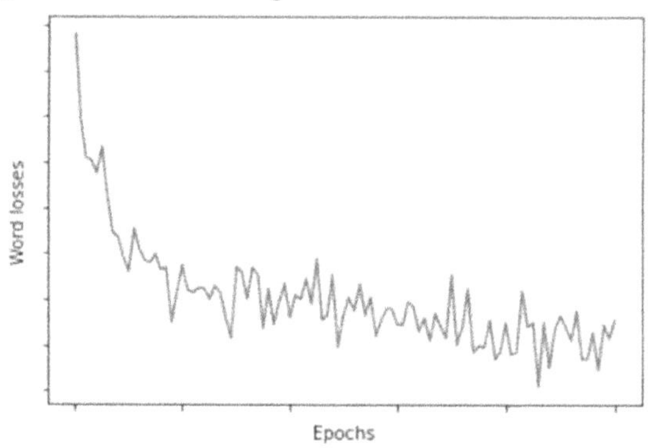

The combined word loss graph tracks the sum of the two-word losses (L_w^0, L_w^1) along with the contrastive learning loss (L_c) across epochs. Initially, there is a significant drop in Figure 9, indicating the encounter of new words. Subsequently, the graph exhibits a jagged pattern, indicating the fine-tuning process of the contrastive learning loss as similar words are encountered in the dataset captions. This pattern persists throughout the 100 epochs, suggesting ongoing fine-tuning. However, stabilization towards the end may occur with further training.

Similar to Figure 9, Figure 10 illustrates the combined sentence loss graph, tracking the sum of the two sentence losses (L_s^0, L_s^1) along with the contrastive learning loss (L_c) across epochs. In Figure 10, an initial sharp decline occurs, indicating the encounter of new sentences. Subsequently, a jagged pattern emerges in the later parts of the graph, signifying the ongoing process of contrastive learning as similar sentences are encountered. This pattern persists throughout the 100 epochs, indicating continuous fine-tuning. Further stabilization may be achieved with additional training, similar to the previous figure.

Figure 10. Tracking sentence loss versus epochs

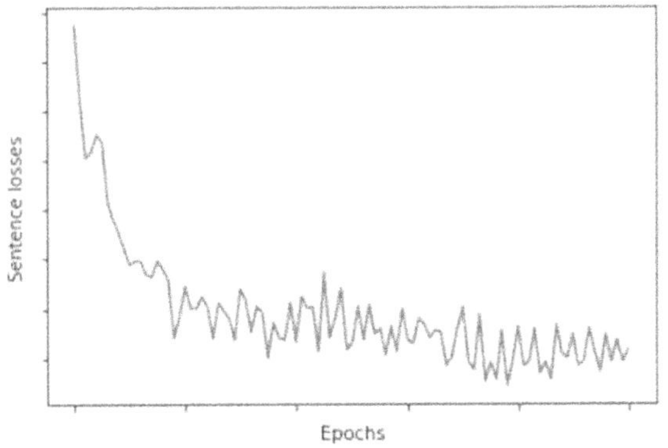

Figure 11. Tracking learning rate of encoders versus epochs

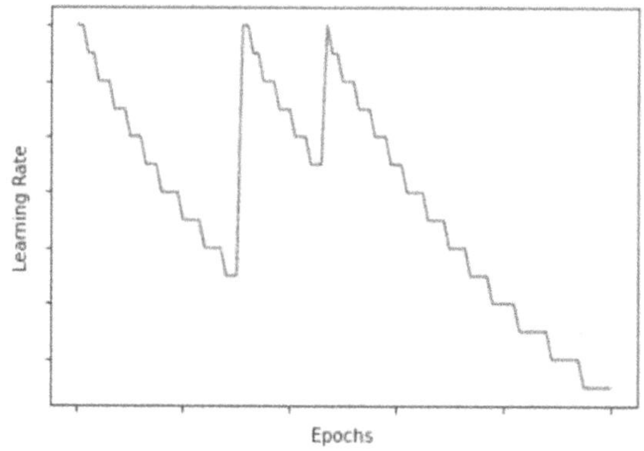

Both the text and image encoders are trained with the same learning rate, and their learning rate patterns are monitored over the epochs. Figure 11 depicts an initial decrease in the learning rate. However, during the middle of the training process, the momentum of the learning rate is adjusted to facilitate fine-tuning. Subsequently, the learning rate remains relatively stable for the remainder of the training, characterized by a gradual step decrease.

As depicted in Figure 12, the trends observed in the generator losses mirror those of the word and sentence losses, owing to the application of contrastive learning to the GAN. The generator is trained for 100 epochs, during which the type of samples evaluated are the fake generated images from the generator. An initial sharp decrease in the losses signifies the discovery of new samples, while the jagged pattern in the latter half indicates the discovery of similar samples. This pattern suggests that a fine-tuning stage has been reached in the training process.

Figure 12. Tracking generator loss versus epochs

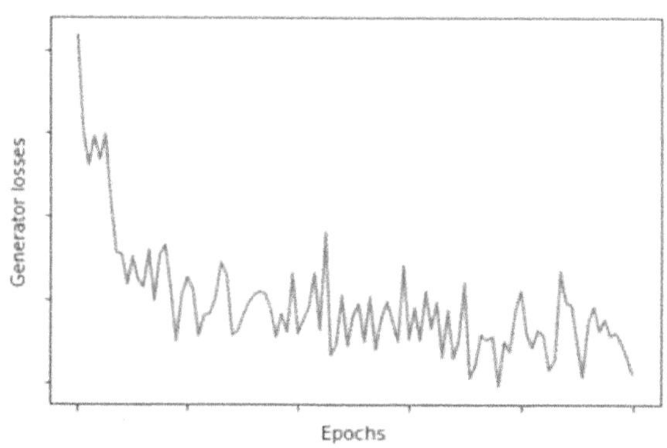

Figure 13. Tracking discriminator loss versus epochs

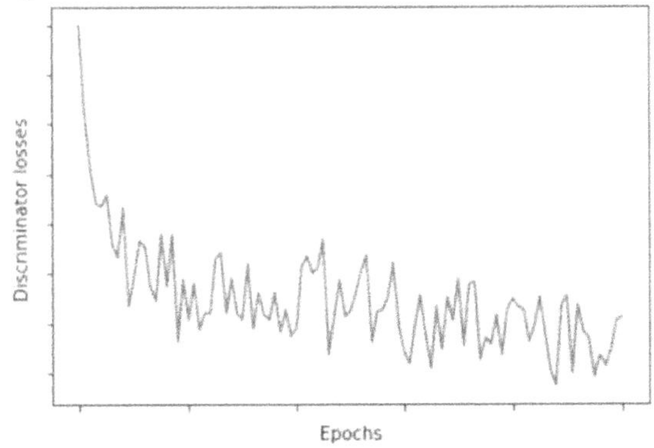

The discriminator losses illustrated in Figure 13, exhibit a similar trend to that of the generator losses, reflecting the evaluation of fake and real samples produced by the generator. Since both components of the GAN are trained concurrently, the discriminator undergoes training for 100 epochs, with contrastive learning applied to it as well. During training, the discriminator assesses the authenticity of the samples generated by the generator, categorizing them as real or fake. Similar to Figure 13, an initial significant decrease in losses indicates a higher number of fake samples being classified, while the jagged pattern observed in the latter half suggests challenges in classifying both fake and real samples. This trend implies that as training progresses, the discriminator faces difficulty in distinguishing between real and fake samples.

Table 2. Loss calculated for various epochs – G denotes Generator and D denotes Discriminator

Model	Epoch 0	Epoch 50	Epoch 100
DAMSM Loss	3.76	2.53	2.51
AttnGAN Loss	G 4.51 D 36.38	G 2.20 D 28.31	G 1.83 D 22.57

The losses are documented across three stages of epochs, as indicated in the Table 2, and monitored in the preceding Figures 9, 10, 11, 12 and 13. A general reduction in the loss is evident in both models; however, the decline becomes less pronounced in the final 50 epochs, suggesting that the model training is approaching a saturation point. The encoders and GAN models underwent testing across diverse test cases, including those absent in the dataset, and were further evaluated on a validation set. The specific test cases have been systematically documented in tabular format.

Two encoders were subject to testing: text and image encoders as shown in Figure 14. The text encoders receive a text caption as input and output a PyTorch tensor, while the image encoders accept an image input and also produce a PyTorch tensor. Both encoders are anticipated to yield encoded tensors comprising features extracted from the text captions and images, respectively.

Text Encoders

The proposed model surpasses similar models when evaluated against performance metrics.

Figure 14. GAN Test cases

S.No	Input Caption	Final Output	Attention Maps
1.	this bird is completely red with black wings and a pointy beak	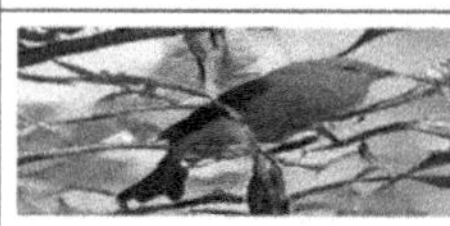	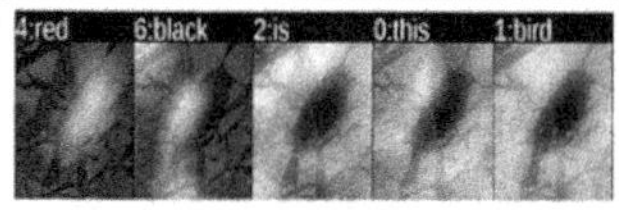
2.	this bird has a red, blue and green tail with a black beak	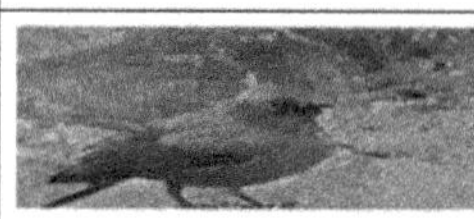	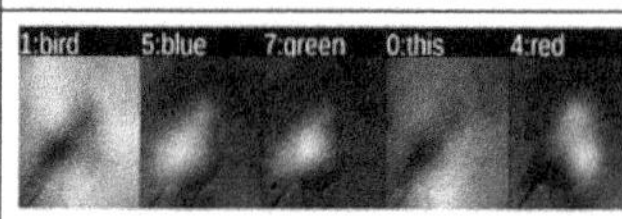
3.	this bird has a blue crown with white throat and brown secondaries		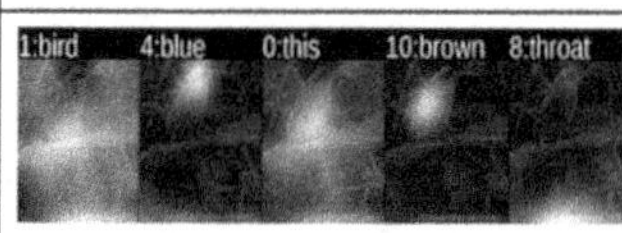
4.	a small colourful bird that contains bright blue feathers covering most of its body except for on its tail	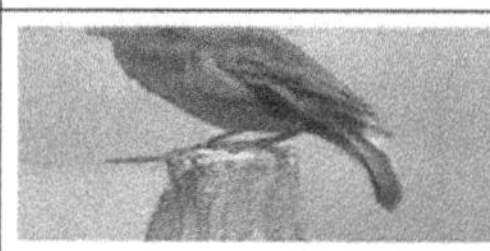	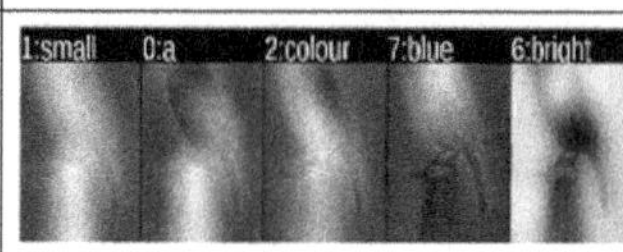
5.	a yellow bird with a grey back, yellow tarsus and brown throat, nape with a black face		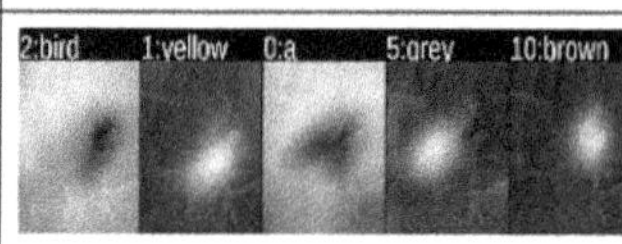

In comparison with the initial attention-driven model AttnGAN (Xu et al., 2019), which lacks contrastive learning, word update, or semantic distillation modules, the proposed model exhibits improvements of +0.35 in Inception Score and a reduced FID score by 6.40 units as shown in Table 3. Further comparison with subsequent attention-driven models reveals a consistent trend of increasing Inception Score and decreasing FID score. Notably, when contrastive learning is applied to the KT-GAN model, the proposed approach demonstrates a marginal decrease of -0.14 in the Inception Score and an improvement of 0.26 in FID. This underscores the significant impact of contrastive learning on GAN training.

Table 3. Comparative analysis of various models with their respective inception score and fretchet inception distance

Model Used	Inception Score	Fretchet Inception Distance
AttnGAN	4.36	23.98
AttnGAN + AATM	4.74	20.40
AttnGAN + SDM	4.76	18.21
KT - GAN	4.85	17.32
Alternating AttnGAN (proposed)	4.71	17.58

CONCLUSION AND FUTURE WORK

The Alternating Attention GAN, an extension of KT-GAN [1], improves text-to-image synthesis through contrastive learning integration. Leveraging core modules from AttnGAN [2], KT-GAN introduced a word feature update mechanism. Contrastive learning enhances image-caption consistency by refining GAN and DAMSM results. This iterative refinement process enhances overall image quality. As GANs evolve, enhancements like those proposed for KT-GAN signify progress. Future advancements will likely explore additional techniques to improve text-to-image synthesis models. Researchers aim to develop efficient GAN models capable of generating high-quality images from textual descriptions.

REFERENCES

Denton, E., Chintala, S., Szlam, A., & Fergus, R. (2015). Deep generative image models using a Laplacian pyramid of adversarial networks. *arXiv (Cornell University), 28*, 1486–1494. https://arxiv.org/pdf/1506.05751

Gregor, K., Danihelka, I., Graves, A., Rezende, D., & Wierstra, D. (2015). *DRAW: A Recurrent Neural Network For Image Generation.*, https://ece.duke.edu/~lcarin/Zhe10.2.2015.pdf

Li, W., Zhang, P., Zhang, L., Huang, Q., He, X., Lyu, S., & Gao, J. (2019). *Object-Driven Text-To-Image synthesis via adversarial training*. https://api.semanticscholar.org/CorpusID:67855413

Mansimov, E., Parisotto, E., Ba, J., & Salakhutdinov, R. (2015b). *Generating Images from Captions with Attention*. https://api.semanticscholar.org/CorpusID:9996719

Nguyen, A., Clune, J., Bengio, Y., Dosovitskiy, A., & Yosinski, J. (2017). *Plug & Play Generative Networks: Conditional Iterative Generation of Images in Latent Space*. https://doi.org/DOI: 10.1109/CVPR.2017.374

Reed, S., Akata, Z., Yan, X., Logeswaran, L., Schiele, B., & Lee, H. (2016, June). Generative adversarial text to image synthesis. In *International conference on machine learning* (pp. 1060-1069). PMLR.

Reed, S., Oord, A., Kalchbrenner, N., Colmenarejo, S. G., Wang, Z., Chen, Y., & Freitas, N. (2017, July). Parallel multiscale autoregressive density estimation. In *International conference on machine learning* (pp. 2912-2921). PMLR.

Tan, H., Liu, X., Li, X., Zhang, Y., & Yin, B. (2019). *Semantics-Enhanced Adversarial Nets for Text-to-Image Synthesis*. https://doi.org/DOI: 10.1109/ICCV.2019.01060

Tan, H., Liu, X., Liu, M., Yin, B., & Li, X. (2021). KT-GAN: Knowledge-Transfer Generative Adversarial Network for Text-to-Image synthesis. *IEEE Transactions on Image Processing, 30*, 1275–1290. DOI: 10.1109/TIP.2020.3026728 PMID: 33001801

Van Den Oord, A., Kalchbrenner, N., Espeholt, L., Kavukcuoglu, K., Vinyals, O., & Graves, A. (2016). *Conditional Image Generation with PixelCNN Decoders*. https://api.semanticscholar.org/CorpusID:14989939

Vaswani, A., Shazeer, N., Parmar, N., Uszkoreit, J., Jones, L., Gomez, A. N., Kaiser, L., & Polosukhin, I. (2017). Attention is All you Need. *arXiv (Cornell University), 30*, 5998–6008. https://arxiv.org/pdf/1706.03762v5

Xu, T., Zhang, P., Huang, Q., Zhang, H., Gan, Z., Huang, X., & He, X. (2018). Attngan: Fine-grained text to image generation with attentional generative adversarial networks. In *Proceedings of the IEEE conference on computer vision and pattern recognition* (pp. 1316-1324).

Zhang, H., Xu, T., & Li, H. (2017). *StackGAN: Text to Photo-Realistic Image Synthesis with Stacked Generative Adversarial Networks*. https://doi.org/DOI: 10.1109/ICCV.2017.629

Zhang, H., Xu, T., Li, H., Zhang, S., Wang, X., Huang, X., & Metaxas, D. N. (2019). StackGAN++: Realistic Image Synthesis with Stacked Generative Adversarial Networks. *IEEE Transactions on Pattern Analysis and Machine Intelligence*, *41*(8), 1947–1962. DOI: 10.1109/TPAMI.2018.2856256 PMID: 30010548

Zhu, Z., Huang, T., Xu, M., Shi, B., Cheng, W., & Bai, X. (2021). Progressive and aligned pose attention transfer for person image generation. *IEEE Transactions on Pattern Analysis and Machine Intelligence*, *1*, 1. Advance online publication. DOI: 10.1109/TPAMI.2021.3068236 PMID: 33755557

Chapter 10
Application of Generative Adversarial Networks in Brahmi Script Recognition

Neha Gautam
G.H. Raisoni College of Engineering, India

Pranali Dhawas
https://orcid.org/0009-0003-4276-2310
G.H. Raisoni College of Engineering, India

Minakshi Ashok Ramteke
https://orcid.org/0009-0009-3889-0954
G.H. Raisoni College of Engineering, India

Mangala Madankar
https://orcid.org/0000-0001-9130-0161
G.H. Raisoni College of Engineering, India

ABSTRACT

This chapter examines the application of Generative Adversarial Networks (GANs) for the recognition and preservation of the Brahmi script, an essential component of cultural and historical heritage. It covers the script's significance, the challenges of ancient script recognition, and introduces GANs, including DCGAN, CycleGAN, and StyleGAN, highlighting their strengths in generating synthetic data, improving image quality, and pattern recognition. The chapter details the methodology of data collection, GAN architecture modifications, and training processes, emphasizing performance metrics like Structural Similarity Index (SSIM), Peak Signal-to-Noise Ratio (PSNR), accuracy, precision, recall, and F1-score. The findings show StyleGANs excel in image quality while DCGANs perform better in accuracy and

DOI: 10.4018/979-8-3693-7152-7.ch010

precision. Practical applications discussed include enhancing historical research, creating digital archives, and developing educational tools. The chapter concludes with future research directions for further improvements in script recognition and preservation.

1 INTRODUCTION

Generative Adversarial Networks (GANs) are an emerging technique used to generate synthetic data, which can help enhance machine learning models by providing diverse, abundant, and cost-effective training examples. This helps overcome the limitations of real data scarcity, protects privacy, and allows the simulation of various scenarios and conditions, thereby improving model accuracy and robustness. GANs are highly popular today and are used in various domains such as character image generation of different scripts, data augmentation, medical imaging, and video synthesis.

Brahmi script is an ancient script that was very popular during the 3rd century BCE in the Indian subcontinent, and 198 South and Southeast Asian scripts were generated from Brahmi script. However, the recognition and interpretation of Brahmi inscriptions pose significant challenges due to their age, the wear and tear they have undergone, and the complexity of their characters.

Script recognition plays a pivotal role in preserving cultural heritage, serving as a bridge between the past and the present by enabling the decipherment and interpretation of ancient texts. Scripts like Brahmi, which date back to the mid-1st millennium BCE, are repositories of historical, cultural, and linguistic information. Recognizing and understanding these scripts is essential for reconstructing the history and societal contexts of ancient civilizations. This process not only brings to light the knowledge and practices of the past but also helps maintain a continuous cultural narrative that is vital for the identity and heritage of present and future generations.

Earlier, traditional approaches such as data augmentation were used for script recognition. However, traditional methods to generate new samples often struggled with obstacles like the accurate decipherment of Brahmi. In the current time, GANs, a sophisticated AI technique known for its ability to generate and distinguish highly accurate images, offer a promising solution to these challenges. GANs represent a groundbreaking intersection of ancient heritage and modern artificial intelligence to overcome these issues.

GANs consist of two neural networks: a generator and a discriminator, which are trained simultaneously through adversarial processes. The generator creates synthetic images based on the input data, while the discriminator evaluates these synthetic images against real data, learning to distinguish between authentic and generated ones. This iterative process enables GANs to produce highly realistic and nuanced samples as outputs. GANs can enhance and reconstruct damaged or incomplete inscriptions, generate synthetic training data to improve recognition models, and distinguish subtle variations in script forms that traditional methods might overlook. This innovative application not only advances the field of optical character recognition (OCR) but also contributes to the preservation and study of ancient texts.

The implications of employing GANs for Brahmi script recognition extend beyond technological innovation. GANs facilitate a deeper understanding of historical and cultural contexts embedded within these texts by improving the accuracy and efficiency of deciphering ancient inscriptions. Recognizing and interpreting Brahmi inscriptions with good accuracy can unlock a wealth of information about early Indian civilization, including its languages, governance, trade, and religious practices. Furthermore, the use of GANs in this domain underscores the potential of AI to contribute to humanities research, demonstrating how advanced machine learning techniques can bridge the gap between technology and historical scholarship.

The accurate recognition of scripts such as Brahmi is crucial for safeguarding the intellectual legacy of ancient civilizations. These scripts contain a wealth of information, including administrative records, religious texts, scientific treatises, and literary works. By decoding these inscriptions, researchers can gain insights into the governance systems, trade practices, religious beliefs, and daily life of ancient societies. This knowledge enhances the understanding of historical developments and contributes to a comprehensive view of human civilization's progress. Without script recognition, much of this valuable information would remain inaccessible, leading to gaps in the historical and cultural knowledge.

Moreover, script recognition fosters the preservation of linguistic diversity, which is a key aspect of cultural heritage. Many modern languages and scripts have evolved from ancient writing systems like Brahmi. By studying and recognizing these ancient scripts, linguists can trace the development and transformation of languages over time. This helps preserve linguistic roots and enrich the cultural tapestry of contemporary societies. Additionally, understanding the evolution of scripts provides insights into the migration, interaction, and integration of different cultural groups, highlighting the interconnectedness of human history.

Next to that, script recognition also has significant implications for educational and scholarly research. Ancient scripts serve as primary sources for historians, archaeologists, and linguists, offering direct evidence of past events, societal structures, and intellectual achievements. Recognizing these scripts accurately allows scholars to conduct detailed analyses, make informed interpretations, and generate new knowledge. This scholarly work, in turn, contributes to educational curricula, enriching the content taught in schools and universities and fostering a deeper appreciation of cultural heritage among students and the general public.

In the context of digital preservation, script recognition technologies, such as those involving GANs, play an increasingly important role. These technologies enable the digitization and restoration of ancient texts, making them accessible to a global audience and protecting them from physical degradation. Digital archives of recognized and deciphered scripts ensure that cultural heritage is preserved for future generations, allowing continuous access to and engagement with historical documents.

In summary, the recognition of scripts like Brahmi is essential for preserving cultural heritage. It provides invaluable insights into ancient civilizations, maintains linguistic diversity, supports scholarly research, and leverages modern technology for digital preservation. Through these efforts, script recognition ensures that the rich cultural and historical legacy of the past continues to inform and inspire the present and future.

2 OVERVIEW OF BRAHMI SCRIPT

Brahmi script is an ancient script. Brahmi script was used from the 3rd BCE (Before the Common Era) to the 6th century (CE) (Seland, 2014; Singh, 2008). It is believed to have been used for the first time in the edicts of Emperor Ashoka, which were inscribed on rocks and pillars across his vast empire. These inscriptions are not only significant for their historical content, documenting Ashoka's conversion to Buddhism and his efforts to promote moral and ethical governance, but also for their linguistic importance. Similarly, this script was used to write various religious books on Buddhism and Jainism culture (Salomon, 1998; Subba, 2008). Brahmi is recognized as the progenitor of most of the scripts used in South and Southeast Asia today, making it a cornerstone in the history of writing in the region. Brahmi script has 10 vowels and 33 consonants (Figure 1).

Figure 1. Characters and vowels of the Brahmi script (Roy & Mandal, 2016)

The evolution of the Brahmi script is a testament to its adaptability and influence. Initially used for Prakrit and Sanskrit languages, Brahmi underwent significant modifications to accommodate the phonetic and structural requirements of various regional languages. This adaptability led to the development of numerous derivative scripts, including Devanagari, Tamil, Kannada, Telugu, Bengali, and others. Each of these scripts preserved the core principles of Brahmi while evolving distinct features that catered to the linguistic needs of their respective languages. This evolutionary process highlights Brahmi's role in shaping the linguistic landscape of the Indian subcontinent, influencing not just written communication but also educational and literary traditions.

The significance of the Brahmi script extends beyond its historical and linguistic impact. Its inscriptions provide a valuable window into the cultural, social, and political life of ancient India. They record a wide range of information, from royal decrees and religious texts to commercial transactions and legal documents. This diversity of content has helped historians and archaeologists reconstruct the socio-economic conditions, religious practices, and administrative systems of ancient Indian societies. Moreover, the widespread use of Brahmi inscriptions in religious contexts, particularly in Buddhism and Jainism, underscores its role in the dissemination of religious and philosophical ideas across Asia, influencing cultures far beyond the Indian subcontinent.

In modern times, the study of Brahmi has not only been crucial for understanding ancient Indian history but also for preserving and revitalizing cultural heritage. Efforts to decipher and interpret Brahmi inscriptions have provided insights into the early stages of Indian civilization and its interactions with other cultures. Additionally, the aesthetic and structural elements of Brahmi have inspired contemporary art, design, and calligraphy, keeping its legacy alive in modern cultural expressions. The Brahmi script thus stands as a symbol of India's rich historical legacy, illustrating the profound impact of ancient writing systems on the development of civilization.

2.1 Properties of Brahmi Script

i. When a consonant and a vowel come together sequence-wise, one or more extra features may be added to that consonant, and new features can be added anywhere. This extra feature is called “matra” and can be seen inside or outside the characters. If one vowel is followed by one consonant and more than one vowel is followed by one consonant, the new character will be called a compound character and complex compound character, respectively (Pal & Chaudhuri, 2004). For example, all vowels in Figure 2 follow a consonant and created compound characters that are presented in Figure 3.
ii. The total number of consonants, vowels, and compound characters in Brahmi script is 33, 10, and 325, respectively. That means the total number of characters in the Brahmi script is 368 (Roy & Mandal, 2016; Siromoney et al., 1983).
iii. The writing style of the Brahmi script is left to right.

Figure 2. Vowels of the Brahmi script (Roy & Mandal, 2016)

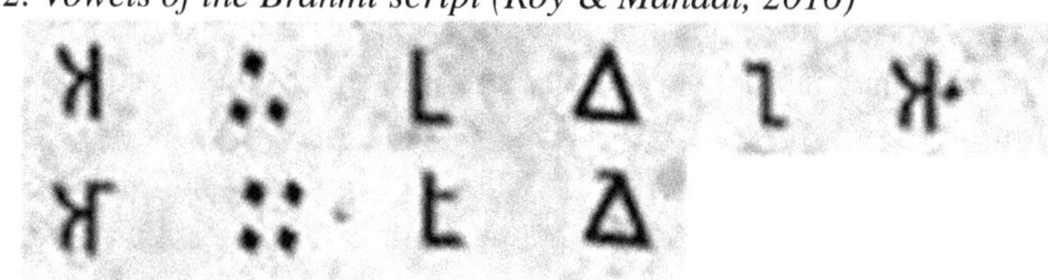

Figure 3. Example of consonant and compound characters (Roy & Mandal, 2016)

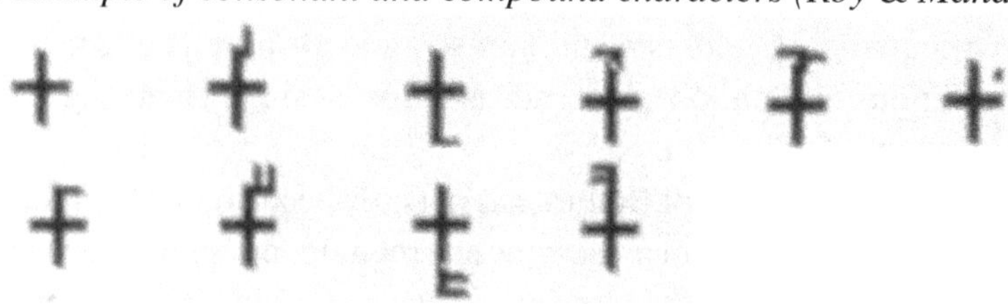

Figure 4 and Figure 5 show an example of Brahmi text where different types of characters, compound characters, and complex compound characters are presented.

Figure 4. Example text in Brahmi script (Rajan, 2010)

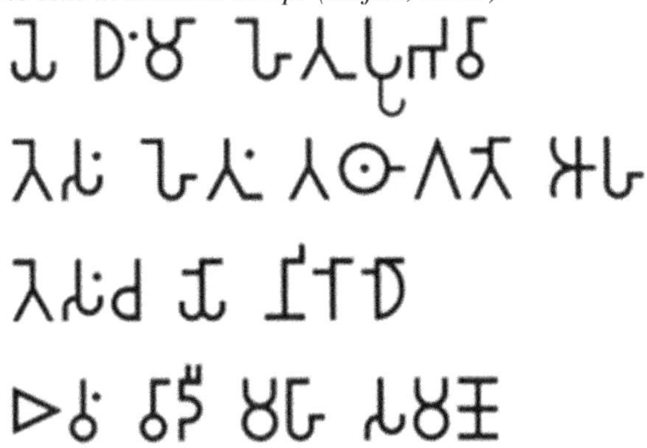

Figure 5. Example of complex compound characters

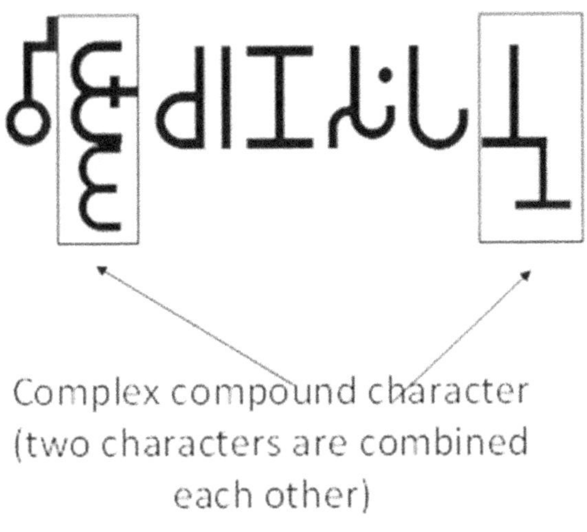

3 IMPORTANCE OF SCRIPT RECOGNITION

This section emphasizes the importance of Brahmi script recognition by highlighting the various challenges associated with character, word, or text recognition, such as deciphering ancient scripts, technological limitations, and a lack of training data. Additionally, it discusses the benefits of preserving ancient scripts, which hold immense cultural and historical significance. These scripts contribute to cultural heritage by offering insights into past civilizations, fostering a sense of identity and continuity, and showcasing the linguistic and cultural diversity of human societies.

3.1 Challenges in Recognizing Ancient Scripts

Recognizing ancient scripts presents numerous challenges, encompassing decipherment complexity, technological limitations, and a lack of training data. Decipherment complexity involves dealing with linguistic variability, incomplete data, and the need for contextual interpretation. Technological limitations arise from poor image quality and variations in handwriting styles. Additionally, the scarcity of annotated datasets and the expertise required for accurate annotation further complicate the process. These challenges collectively hinder the accurate recognition and interpretation of ancient scripts, highlighting the need for advanced methods and technologies.

I. Decipherment Complexity:
    - Linguistic Variability: Ancient scripts like Brahmi often display significant variability in their characters and forms. This is due to their evolution over centuries and the influence of different regional languages and dialects.
    - Incomplete Data: Many ancient inscriptions are fragmentary, with parts of the text eroded or missing. This makes it difficult to reconstruct the complete message accurately.
    - Contextual Interpretation: Understanding the context in which the script was used is crucial. Without sufficient historical, cultural, or archaeological context, interpreting the meaning of the text can be challenging.

II. Technological Limitations:
    - Image Quality: Historical inscriptions are often found on materials that have degraded over time, such as stone, metal, or parchment. Poor image quality due to weathering, wear, or damage complicates the recognition process.

- Handwriting Variations: Variations in handwriting styles across different regions and periods further complicate script recognition. Ancient scripts lack the standardization seen in modern writing systems.

III. Lack of Training Data:
- Limited Corpus: Unlike modern languages, the corpus of ancient scripts like Brahmi is limited. The scarcity of large, annotated datasets makes it challenging to train machine learning models effectively.
- Annotation Challenges: Even when data is available, annotating it accurately requires expertise in paleography and historical linguistics, which are specialized fields.

3.2 Cultural and Historical Significance of Preserving Ancient Scripts

I. Cultural Heritage:
- Identity and Continuity: Ancient scripts are a crucial part of cultural identity and heritage. They provide insights into the lives, beliefs, and practices of past civilizations, fostering a sense of continuity and connection with ancestors.
- Diversity and Richness: The preservation and study of ancient scripts like Brahmi highlight the linguistic and cultural diversity of human societies. This diversity is a testament to the richness of human expression and intellectual achievement over millennia.

II. Historical Understanding
- Chronological Insights: Ancient inscriptions often contain valuable historical data, such as dates, events, and names of rulers. Deciphering these scripts can provide accurate timelines and chronologies of historical events.
- Sociopolitical Context: Scripts like Brahmi offer insights into the sociopolitical structures of ancient societies. They reveal information about governance, law, trade, and social hierarchies, contributing to a comprehensive understanding of historical contexts.

III. Linguistic Evolution:
- Language Development: Studying ancient scripts helps linguists trace the development and evolution of languages. It reveals how languages transform over time and how modern languages are connected to their ancient predecessors.
- Script Innovation: Brahmi is one of the earliest known scripts in the Indian subcontinent and is considered the precursor too many modern

Indian scripts. Understanding its structure and evolution can shed light on the innovation and adaptation processes in script development.

IV. Educational and Scholarly Value:
- Academic Research: Preserving and studying ancient scripts like Brahmi is crucial for academic research in fields such as history, archaeology, linguistics, and anthropology. It provides primary sources for scholarly investigation and discourse.
- Educational Resource: These scripts serve as valuable educational resources, helping students and researchers learn about ancient civilizations, their languages, and their contributions to human knowledge.

4 GENERATIVE ADVERSARIAL NETWORKS (GANS)

Generative Adversarial Networks (GANs) are a groundbreaking class of machine learning models that involve a generator creating synthetic data and a discriminator evaluating its authenticity. Over time, GANs have evolved into notable variants like DCGAN, CycleGAN, and StyleGAN, each excelling in different applications. DCGANs are known for stable training and high-quality image generation, CycleGANs for unpaired image-to-image translation, and StyleGANs for detailed and customizable image synthesis. The strengths and applications of each variant have significantly advanced the capabilities of image generation and transformation.

4.1 Introduction to GANs: Explanation of Generator and Discriminator Roles

Generative Adversarial Networks (GANs) are a class of machine learning frameworks designed by Ian Goodfellow and his colleagues in 2014 (Goodfellow et al., 2014). GAN consist of two neural networks, the generator and the discriminator, which are trained simultaneously through adversarial processes. This innovative architecture enables GANs to generate new, synthetic instances of data that resemble the real data on which they were trained.

I. Roles of the Generator and Discriminator
- **Generator (G):** The role of generator is to create synthetic data that is as realistic as possible. It takes a random noise vector z as input and transforms it into a data sample $G(z)$. The goal of the generator is to produce data that the discriminator cannot distinguish from real data. This is a convolutional neural network (CNN) that takes random noise as input and generates a synthetic image. The generator typically includes

several layers to progressively upsample the image resolution and add details. It takes a random noise vector z as input and outputs synthetic data $G(z)$.

$G{:}z \rightarrow G(z)$

- **Discriminator (D):** The role of discriminator is to distinguish between real and synthetic data. It takes a data sample as input and outputs a probability score, *D(x)*, indicating whether the sample is real (close to 1) or fake (close to 0). The discriminator is trained to maximize its accuracy in differentiating real data samples from those generated by the generator. This CNN takes an image as input and classifies it as either real or fake. The discriminator consists of several layers to downsample the image resolution and extract features. It evaluates both real data xxx and synthetic data *G(z)*, outputting a probability score D(x) or D(G(z)) that indicates the likelihood of the input being real.

D:x→D(x)

II. Training Process

The training process of GANs involves a minimax game between the generator and the discriminator. The objective function for this game can be described by the following equation:

$$\min_G \max_D V(D,G) = \mathbb{E}_{x \sim p_{data}(x)}[logD(x)] + \mathbb{E}_{z \sim p_z(z)}[\log(1 - D(G(z)))]$$

- **Objective for the Discriminator:** The discriminator aims to maximize the probability of correctly classifying real and generated data, thus maximizing $logD(x)$ for real samples and $\log(1-D(G(z)))$ for fake samples.
- **Objective for the Generator:** The generator aims to minimize $\log(1-D(G(z)))$, which corresponds to generating data that the discriminator incorrectly classifies as real.

III. Training Dynamics

1. **Initialization:** Both the generator and discriminator are initialized with random weights.
2. **Discriminator Update:** The discriminator is trained using real data samples *x* and synthetic data samples *G(z)*. It adjusts its weights to maximize its classification accuracy.

3. **Generator Update:** The generator is trained to produce data samples that maximize the discriminator's probability of classifying them as real. This is done by adjusting its weights to minimize the discriminator's ability to distinguish between real and fake data.
4. **Iterative Process:** Steps 2 and 3 are repeated iteratively, with the generator continuously improving its output to fool the discriminator, and the discriminator getting better at detecting fake data.

This adversarial training process can be visualized as follows:

Figure 6. GAN training process

Here,

- The generator network *(G)* generates synthetic data from random noise.
- The discriminator network *(D)* evaluates the data, trying to distinguish between real and synthetic data.
- Both networks are trained simultaneously, improving the performance of each other through their adversarial relationship.

4.2 Evolution and Notable Variants: DCGAN, CycleGAN, StyleGAN

Generative Adversarial Networks (GANs) have undergone significant evolution since their introduction, with several notable variants developed to address specific challenges and enhance the capabilities of the original GAN architecture. Each variant introduces modifications to the basic GAN framework, leading to advancements in generating more realistic images, preserving image quality, and performing complex transformations. Here's an overview of some prominent GAN variants along with their key equations.

4.2.1 Deep Convolutional GAN (DCGAN)

Deep Convolutional Generative Adversarial Networks (DCGANs) are a sophisticated type of generative model that leverage adversarial training to create realistic and diverse images. Picture two competitors: one (the generator) continuously attempts to create forgeries, while the other (the discriminator), a skilled detective, strives to identify the fakes. This perpetual rivalry enhances both their abilities, eventually enabling the generator to produce images that can even deceive the sharp-eyed discriminator.

Figure 7. Flow diagram of DCGAN

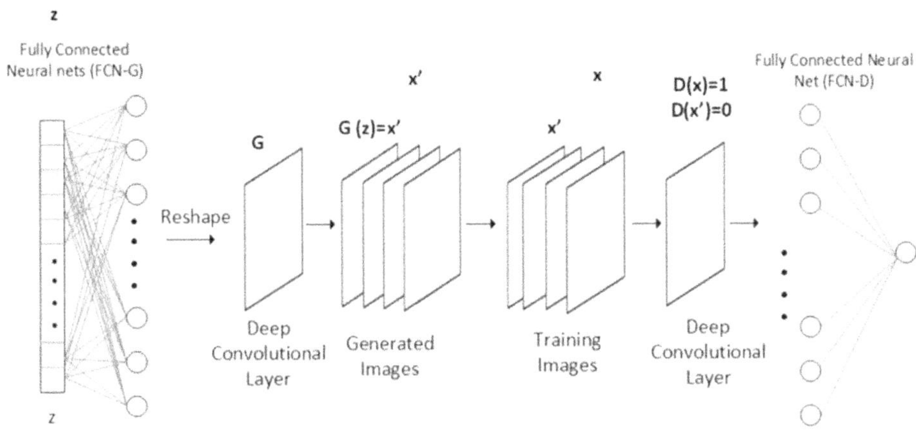

Here's an overview of the DCGAN process:

The mathematical framework for DCGANs involves defining the generator and discriminator networks and their respective loss functions for training. Let the generator be denoted as G and the discriminator as D.

Generator: This is a convolutional neural network (CNN) that takes random noise as input and generates a synthetic image. The generator typically includes several layers to progressively upsample the image resolution and add details. It takes a random noise vector z as input and outputs synthetic data G(z).

$G{:}z \rightarrow G(z)$

Discriminator: This CNN takes an image as input and classifies it as either real or fake. The discriminator consists of several layers to downsample the image resolution and extract features. It evaluates both real data xxx and synthetic data

G(z), outputting a probability score D(x) or D(G(z)) that indicates the likelihood of the input being real.

$$D{:}x \rightarrow D(x)$$

$$D{:}G(z) \rightarrow D(G(x))$$

Loss functions: Training involves distinct loss functions for the generator and discriminator in an adversarial manner. The generator's loss function encourages the creation of images that deceive the discriminator, while the discriminator's loss function aims to accurately distinguish between real and synthetic images. Both networks are trained using binary cross-entropy loss functions, minimizing the difference between predicted probabilities and actual labels (1 for real, 0 for fake).

Generator Loss

$$L_G = -\frac{1}{m}\sum_{i=1}^{m} logD(G(z^{(i)}))$$

Discriminator Loss

$$L_D = -\frac{1}{m}\sum_{i=1}^{m}\left[D(x^{(i)}) + \log\left(1 - D(G(z^{(i)}))\right]$$

where mmm is the batch size, $z^{(i)}$ is the random noise vector for the ith sample, and $x^{(i)}$ is the ith real data sample.

4.2.2 CycleGAN

Cycle-Consistent Generative Adversarial Networks, or CycleGANs, are a type of generative model designed for image-to-image translation tasks where paired training data is not available. Imagine two artists, each trying to mimic the other's style: one artist (the generator) creates artwork in a new style, while the other artist (the discriminator) evaluates how convincing the new style is. This adversarial process, coupled with cycle consistency, ensures that the generated images retain their original content when translated back to their initial form.

Figure 8. Flow diagram of CycleGANs

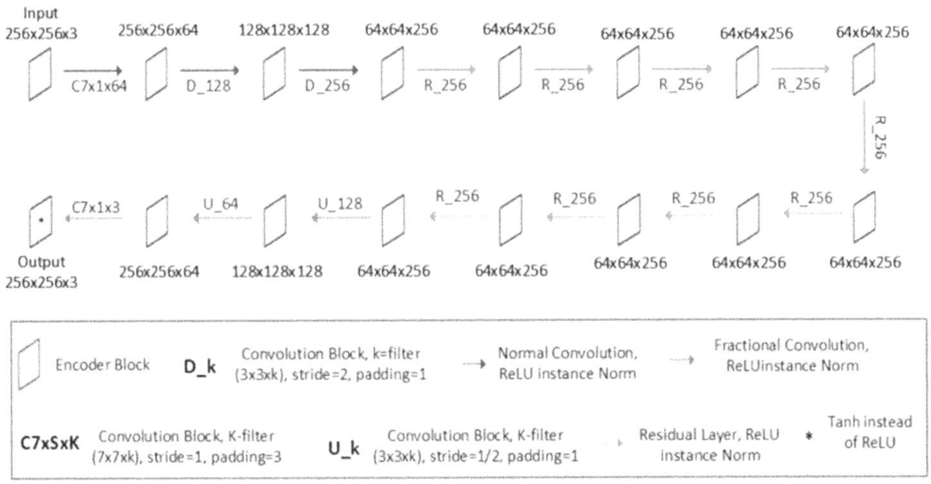

Here's a breakdown of the CycleGAN process:

- **Generators:** CycleGAN employs two generators, G and F. G translates images from domain X to domain Y, while F translates images from domain Y to domain X. The generators aim to create images that are indistinguishable from the target domain.

  $G: X \rightarrow Y$

  $F: Y \rightarrow X$

- **Discriminators:** Similarly, two discriminators, D_X and D_Y, are used. D_X distinguishes between real images in domain X and those generated by F, while D_Y does the same for domain Y and images generated by G.

$D_X: X \rightarrow [0,1]$

$D_Y: Y \rightarrow [0,1]$

- CycleGAN Loss Functions

CycleGAN uses two sets of GAN loss functions along with cycle consistency loss to ensure that the image translation preserves the original content.

- **Adversarial Loss:**

For each domain, CycleGAN uses an adversarial loss to encourage the generators to produce images that are indistinguishable from real images in the target domain.

$$L_{GAN}(G, D_Y, X, Y) = \mathbb{E}_{y \sim p_{data}(y)}[log\, D_Y(y)] + \mathbb{E}_{x \sim p_{data}(x)}[log\, (1 - D_Y(G(x)))]$$

$$L_{GAN}(G, D_X, Y, X) = \mathbb{E}_{x \sim p_{data}(x)}[log\, D_X(x)] + \mathbb{E}_{y \sim p_{data}(y)}[log\, (1 - D_X(F(y)))]$$

- **Cycle Consistency Loss:**

Cycle consistency loss ensures that an image translated to the other domain and back remains unchanged.

$$L_{cycle}(G, F) = \mathbb{E}_{x \sim p_{data}(x)}[\|F(G(x)) - X\|_1] + \mathbb{E}_{y \sim p_{data}(y)}[\|G(F(y)) - y\|_1]$$

where L_{cycle} ensures that the image translations preserve the content.

- **Total Loss:**

The total loss for the generators and discriminators in CycleGAN combines the adversarial and cycle consistency losses.

$$L_{G,F} = L_{GAN}(G, D_Y, X, Y) + L_{GAN}(G, D_X, Y, X) + \lambda \mathscr{L}_{cyc}(G, F)$$

where λ is a weighting factor that balances the importance of the cycle consistency loss relative to the adversarial losses.

Adversarial Loss: The generators and discriminators are trained with adversarial loss. For generator G and discriminator D_Y:

$$L_{GAN}(G, D_Y, X, Y) = \mathbb{E}_{y \sim p_{data}(y)}[log\, D_Y(y)] + \mathbb{E}_{x \sim p_{data}(x)}[\log(1 - D_Y(G(x)))]$$

StyleGAN

Style-Based Generative Adversarial Networks, or StyleGANs, are a variant of GANs that introduce a style-based generator architecture to achieve high levels of control and variation in generated images. Picture a sculptor who not only creates lifelike statues but can also effortlessly change the style, texture, and details of each creation, producing a wide range of artistic variations.

Figure 9. Flow diagram of StyleGANs

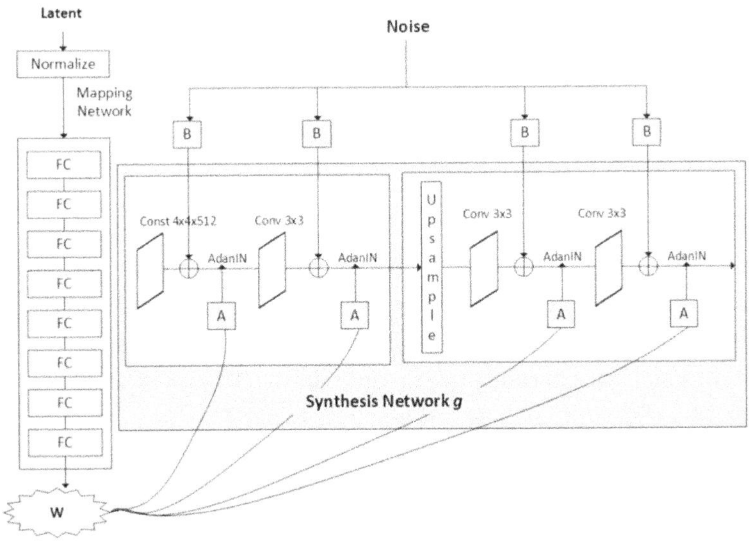

Here's a breakdown of the StyleGAN process:

- **Generator:** The StyleGAN generator consists of two main components: a mapping network and a synthesis network. The mapping network transforms the input latent vector z into an intermediate latent space w, which is then used to modulate the synthesis network at different levels.

$w = M(z) \in$

The synthesis network uses adaptive instance normalization (AdaIN) layers to control the style at each level of the generator, allowing for detailed and hierarchical style manipulation.

$$AdaIN(w,x) = w.\left(\frac{x-\mu(x)}{\sigma(x)}\right) + b$$

- **Discriminator:** The discriminator in StyleGAN operates similarly to other GANs, distinguishing between real and fake images. It processes the images through several convolutional layers to predict the authenticity of the input.

$w: X \rightarrow [0,1]$

- **Loss Functions:** StyleGAN employs standard GAN loss functions, optimizing both the generator and discriminator using binary cross-entropy loss. Additionally, it includes a path length regularization term to ensure smooth image generation and consistent style changes.

4.3 Strengths and Applications of Each Variant

Generative Adversarial Networks (GANs) and their variants, such as DCGANs, CycleGANs, and StyleGANs, each exhibit unique strengths and applications. These GAN variants collectively enhance diverse domains, including computer vision, entertainment, and scientific research, by pushing the boundaries of what can be achieved with synthetic data and image manipulation. Below, the details of strength, application of variants of GAN is discussed further.

4.3.1 Deep Convolutional GAN (DCGAN)

DCGANs, known for their stable training and ability to generate high-quality images, are widely used in tasks like image synthesis, super-resolution, and data augmentation. These strengths and applications make DCGANs highly effective for various image-related tasks, as discussed in further detail.

Strengths:

- **Improved Image Quality**: DCGANs utilize deep convolutional networks in both the generator and discriminator, significantly improving the quality of generated images compared to earlier GAN models.
- **Stability in Training**: The use of deep convolutional layers and batch normalization helps stabilize the training process, reducing issues like mode collapse and instability.
- **High Resolution**: DCGANs can generate high-resolution images with more detailed and coherent structures.

Applications:

- **Image Generation**: DCGANs are widely used for generating realistic images from random noise, making them useful in creative fields like art and design.
- **Data Augmentation**: They can create synthetic data for training other machine learning models, especially when real data is scarce or expensive to obtain.
- **Image Super-Resolution**: DCGANs can be applied to enhance the resolution of images, improving their clarity and detail.

4.3.2 CycleGAN

CycleGANs, renowned for their capability to perform image-to-image translation without requiring paired datasets, excel in applications such as style transfer, object transfiguration, and photo enhancement. These strengths and applications make CycleGANs particularly useful for transforming and enhancing images in diverse and innovative ways, as discussed in further detail.

Strengths:

- **Unpaired Image-to-Image Translation**: CycleGANs excel in translating images between domains without requiring paired examples, which is valuable when paired datasets are not available.
- **Cycle Consistency**: They maintain the structural and semantic integrity of images through cycle consistency loss, ensuring that translations are meaningful and reversible.
- **Flexibility**: CycleGANs can handle complex transformations and generate high-quality images even with diverse and unpaired training data.

Applications:

- **Style Transfer**: CycleGANs can be used to apply artistic styles to images, such as converting photographs into paintings in the style of famous artists.
- **Domain Adaptation**: They are effective in adapting images from one domain to another, such as converting day-time images to night-time or transforming sketches into realistic images.
- **Image Editing**: CycleGANs can perform various image editing tasks, including changing the appearance of objects or altering the background of images.

4.3.2 StyleGAN

StyleGANs stand out for their superior control over image generation through disentangled representations, enabling fine-grained manipulations in facial attribute synthesis, artistic content creation, and high-fidelity image generation. These strengths and applications allow StyleGANs to produce highly detailed and customizable images, making them invaluable for creative and realistic image generation, as discussed in further detail.

Strengths:

- **Fine-Grained Control**: StyleGANs offer unprecedented control over the generated images by separating high-level features from low-level details through a style-based architecture.
- **High Fidelity and Diversity**: They produce high-resolution images with fine details and diverse attributes, making them ideal for generating photorealistic images.
- **Adaptive Instance Normalization**: The use of AdaIN allows StyleGANs to control the style of images at different levels of abstraction, enhancing their versatility and realism.

Applications:

- **Realistic Image Synthesis**: StyleGANs are widely used for generating highly realistic human faces and other complex images, which can be applied in entertainment, gaming, and virtual reality.
- **Data Augmentation**: They can create diverse synthetic datasets for training machine learning models, particularly in scenarios where real data is limited or difficult to obtain.
- **Content Creation**: StyleGANs enable the creation of novel and diverse content, such as generating unique characters for games, avatars for virtual environments, or new fashion designs.

5 APPLICATION OF GANS IN BRAHMI SCRIPT RECOGNITION

This section focuses on how GANs can be applied to the specific problem of Brahmi script recognition. It will outlines the challenges of recognizing Brahmi script and explains why GANs are well-suited to address these challenges, such as their ability to generate synthetic data and improve image quality. The following major topics will be covered in this section

5.1 Challenges in Brahmi Script Recognition: Variability, Weathering, And Limited Data

Recognizing the ancient Brahmi script presents several challenges due to its inherent variability, the effects of weathering on inscriptions, and the scarcity of available data. Variability in characters across regions and time periods complicates consistent recognition, while the weathering of materials like stone and metal further degrades the clarity of inscriptions. Additionally, the limited corpus of Brahmi text data makes it difficult to develop and train effective recognition models, adding another layer of complexity to the task.

5.1.1 Variability in Script Forms

One of the primary challenges in Brahmi script recognition is the high variability in script forms. Brahmi, being an ancient script, evolved significantly over time and across different regions. This evolution led to the development of various regional and stylistic variants, each with unique characters and forms. The diversity in Brahmi inscriptions, including variations in stroke styles, character shapes, and writing orientations, complicates the recognition process. These variations can occur within the same inscription or across different inscriptions, making it difficult for recognition algorithms to accurately identify and classify the characters.

To address this challenge, advanced techniques in machine learning and pattern recognition are required. Models need to be trained on a broad range of Brahmi script examples to learn the nuances and variations. Developing robust algorithms that can generalize across different forms of Brahmi while maintaining high accuracy is a key challenge in script recognition.

5.5.2 Weathering and Degradation of Inscriptions

The physical weathering and degradation of Brahmi inscriptions pose another significant challenge. Many Brahmi inscriptions are found on ancient stone tablets, pillars, and other weathered surfaces. Over centuries, these inscriptions have been subjected to environmental factors such as erosion, weathering, and pollution. This deterioration can result in incomplete or partially erased characters, making them difficult to read and interpret.

To mitigate this issue, digital imaging and restoration techniques are often employed. High-resolution scans, multispectral imaging, and other advanced methods can help reveal faded inscriptions and reconstruct damaged characters. However, the effectiveness of these techniques depends on the quality of the original inscriptions and the extent of their degradation. Accurate recognition of weathered and fragmented inscriptions requires sophisticated image processing algorithms and expertise in historical document restoration.

5.5.3 Limited Data Availability

The limited availability of annotated data is a critical challenge in Brahmi script recognition. Creating a comprehensive dataset of Brahmi inscriptions, especially with annotations for training machine learning models, is a complex and resource-intensive task. Many existing inscriptions are in fragmented or incomplete forms, and the process of manually transcribing and annotating these texts is labor-intensive and requires specialized knowledge.

To overcome this limitation, researchers are exploring various approaches, such as generating synthetic data, leveraging transfer learning, and crowdsourcing annotations. Synthetic data can be created by simulating Brahmi script variations and degradation patterns, providing additional training examples for models. Transfer learning techniques involve using pre-trained models on related tasks and fine-tuning them for Brahmi script recognition. Crowdsourcing can also help in gathering and annotating data by involving a larger community of experts and enthusiasts.

In the conclusion, Brahmi script recognition faces several challenges, including the variability in script forms, weathering and degradation of inscriptions, and limited data availability. Addressing these challenges requires a combination of advanced machine learning techniques, digital imaging methods, and innovative approaches to data collection and annotation. Overcoming these obstacles is crucial for accurately deciphering and preserving the rich historical and cultural heritage encoded in Brahmi inscriptions.

5.2 Advantages of Using GANs: Synthetic data generation, image quality improvement, and enhanced pattern recognition

Generative Adversarial Networks (GANs) offer significant advantages in various domains, particularly in synthetic data generation, improving image quality, and enhancing pattern recognition. By generating realistic synthetic data, GANs provide valuable resources for training models when real data is scarce or sensitive. Their ability to produce high-quality images is utilized in tasks like super-resolution and image restoration, while their advanced pattern recognition capabilities enhance the accuracy and effectiveness of models in fields such as medical imaging and computer vision.

I. Synthetic Data Generation

One of the most significant advantages of using Generative Adversarial Networks (GANs) is their capability for synthetic data generation. GANs can create high-quality synthetic data that mimics real-world data distributions, which is especially valuable when real data is scarce, incomplete, or difficult to obtain. This capability is particularly beneficial in domains such as medical imaging, where acquiring annotated samples can be challenging due to privacy concerns and the need for expert involvement.

In the context of Brahmi script recognition, GANs can generate synthetic examples of Brahmi characters and inscriptions. These synthetic samples can help augment existing datasets, providing more diverse and comprehensive training data for machine learning models. By simulating variations in script forms, degradation patterns, and writing styles, GANs enhance the robustness of recognition algorithms and improve their generalization capabilities.

II. Image Quality Improvement

GANs are well-known for their ability to enhance image quality through techniques such as super-resolution and de-noising. By learning from high-resolution and low-resolution image pairs, GANs can generate detailed, high-resolution images from lower-resolution inputs. This feature is particularly useful for improving the quality of historical documents, including weathered and degraded Brahmi inscriptions.

For Brahmi script recognition, GANs can be used to reconstruct faded or damaged inscriptions, making the text more legible and easier to analyze. By improving the clarity and detail of historical images, GANs facilitate more accurate recognition and interpretation of ancient scripts. This enhancement helps preserve the historical integrity of the inscriptions while making them more accessible for research and study.

III. Enhanced Pattern Recognition

GANs contribute to enhanced pattern recognition by learning complex data distributions and generating high-quality samples that capture intricate details. The adversarial training process, where the generator and discriminator improve through competition, enables GANs to identify and replicate subtle patterns and features in data. This capability is advantageous for tasks involving fine-grained pattern recognition, such as distinguishing between different forms of Brahmi characters.

In Brahmi script recognition, GANs can aid in identifying and categorizing variations in script forms and styles. By generating diverse examples and training models on these synthetic samples, GANs help improve the accuracy and sensitivity of recognition algorithms. Enhanced pattern recognition facilitates the accurate decoding of ancient scripts, even when dealing with variations and inconsistencies in historical inscriptions.

In the conclusion, the advantages of using GANs in script recognition are substantial. Their ability to generate synthetic data augments training datasets, providing more diverse and comprehensive examples for machine learning models. GANs also enhance image quality, allowing for improved visibility and readability of historical documents. Additionally, their capability for enhanced pattern recognition supports the accurate identification and classification of complex script variations. By leveraging these strengths, GANs significantly contribute to the advancement of script recognition technologies and the preservation of cultural heritage.

6 METHODOLOGY

This section details the research methodology, including how Brahmi script data is collected and prepared, the specific GAN architecture used, and any modifications made to suit the application, and the training process. It provides a step-by-step guide to how the research was conducted. The following major topics will be covered in this section.

o Data Collection and Preparation: Digitizing inscriptions and dataset creation.

In the application of Generative Adversarial Networks (GANs) for Brahmi script recognition, the accuracy and reliability of the model largely depend on the quality and comprehensiveness of the dataset used for training. The process begins with the meticulous digitization of ancient Brahmi inscriptions, followed by the creation of a structured dataset tailored for GAN training.

6.1 Digitizing Inscriptions

The first step involves sourcing high-resolution images of Brahmi script inscriptions from historical records, archaeological findings, and digital archives. These inscriptions are often found on ancient artifacts such as stone tablets, pillars, and manuscripts, with varying degrees of wear and weathering due to their age. To ensure the fidelity of the digitization process, advanced imaging techniques, such as 3D scanning, multispectral imaging, and photogrammetry, can be are employed. However, this is expensive and time taken process. Thus, this study considered a printed text based dataset of Brahmi script (Gautam et al., 2020).

2. Preprocessing and Annotation

Once the inscriptions are digitized, the images undergo a preprocessing phase to enhance clarity and uniformity. This includes noise reduction, contrast adjustment, and normalization of image dimensions. Additionally, each character within the inscriptions is carefully annotated to label the characters and segment the inscriptions into individual glyphs. This manual annotation process is essential for creating a ground truth dataset that can be used to train the GANs effectively.

3. Dataset Creation

The annotated images are then organized into a structured dataset, categorized by character, word, or inscription. Given the inherent variability in Brahmi script due to regional and temporal differences, the dataset is designed to capture a wide range of character forms and styles. To address the challenge of limited data, the dataset is augmented using techniques such as rotation, scaling, and noise addition, which increase the diversity of training samples without compromising the authenticity of the original inscriptions.

The final dataset is split into training, validation, and testing subsets, ensuring that the model is exposed to a comprehensive range of examples during the training phase while reserving unseen samples for evaluation. This structured and well-prepared dataset forms the foundation for training GANs, enabling the accurate recognition and reconstruction of Brahmi script characters.

o GAN Architecture: Specific architecture used, modifications, and customizations.

The architecture of Generative Adversarial Networks (GANs) plays a crucial role in determining the quality and effectiveness of the generated outputs, especially in specialized applications like Brahmi script recognition. This section discusses the specific architectures used, along with modifications and customizations made to enhance performance, focusing on DCGAN, StyleGAN, and CycleGAN.

1. DCGAN (Deep Convolutional GAN)

DCGAN is a popular variant of GAN that leverages deep convolutional networks to generate high-quality images. It is known for its stable training process and the ability to create detailed and realistic images.

- **Text Explanation**: The DCGAN architecture consists of a generator and a discriminator, both of which are convolutional neural networks (CNNs). The generator network uses a series of transposed convolutional layers (also known as deconvolutional layers) to upsample a noise vector into an image. The discriminator, on the other hand, uses convolutional layers to downsample the input image and classify it as real or fake. Key modifications in DCGAN for Brahmi script recognition include adjusting the filter sizes, adding batch normalization layers, and using leaky ReLU activation functions to stabilize the training.

2. StyleGAN

StyleGAN is an advanced GAN variant designed for high-resolution image synthesis, allowing for fine-grained control over the style of generated images at different levels of detail.

- **Text Explanation**: Unlike traditional GANs, StyleGAN introduces a novel generator architecture that separates the latent space into two stages: a mapping network that converts the latent vector into an intermediate space, and a synthesis network that generates the image. This architecture allows for style mixing and fine control over image features, making it particularly useful for generating high-resolution Brahmi script images with varying styles. Customizations for Brahmi script recognition include tuning the style layers to capture the intricate details of different script forms and adjusting the resolution to match the script's characteristics.

3. CycleGAN

CycleGAN is designed for unpaired image-to-image translation tasks, enabling the transformation of images from one domain to another without requiring paired training data.

- **Text Explanation**: The CycleGAN architecture consists of two generators and two discriminators, forming a cycle-consistent adversarial model. For Brahmi script recognition, CycleGAN can be customized to translate degraded or weathered script images into clearer, more legible forms. The cycle consistency loss ensures that translating an image from domain A to B and back to A results in the original image, preserving the structural integrity of the script. Modifications include tuning the cycle consistency weight to ensure high-fidelity translations and adjusting the generator architectures to handle the unique characteristics of Brahmi script images.

o Training Process: Data augmentation, hyperparameter tuning, handling overfitting.

The training process of Generative Adversarial Networks (GANs) for Brahmi script recognition involves several critical steps to ensure that the models generate accurate and diverse outputs. This section outlines the key components of the training process, including data augmentation, hyperparameter tuning, and techniques for handling overfitting, incorporating both textual explanations and code-based examples.

1. Data Augmentation

Data augmentation is crucial for enhancing the training dataset, especially when working with limited or imbalanced data, as is often the case with ancient scripts like Brahmi. Augmentation techniques are employed to increase the diversity of the training data, improving the model's ability to generalize.

- **Text Explanation**: Common data augmentation techniques include rotation, scaling, flipping, and adding noise. These augmentations help simulate the variability in real-world data, such as different handwriting styles, degraded inscriptions, and various orientations of the script. By generating multiple augmented versions of each training sample, the model is exposed to a wider range of input scenarios.

2. Hyperparameter Tuning

Hyperparameter tuning is a critical step in optimizing the performance of GANs. It involves adjusting parameters such as learning rates, batch sizes, and the architecture's depth and width to achieve the best possible results.

- **Text Explanation**: Hyperparameters like the learning rate for both the generator and discriminator, the number of epochs, and the batch size significantly influence the training dynamics. For instance, a learning rate that's too high might cause the model to diverge, while one that's too low might slow down convergence. Batch size also impacts the stability of GAN training, with smaller batches sometimes leading to better generalization.

3. Handling Overfitting

Overfitting is a common challenge in deep learning, where the model performs well on training data but poorly on unseen data. In GANs, overfitting can occur in both the generator and discriminator, leading to poor generalization and a failure to produce realistic outputs.

- **Text Explanation**: Techniques such as dropout, regularization, and early stopping are employed to mitigate overfitting. Dropout randomly deactivates neurons during training, which prevents the model from becoming too dependent on specific pathways. Regularization techniques, like L2 regularization, add a penalty to the loss function to discourage overly complex models. Early stopping monitors the model's performance on a validation set and halts training when performance no longer improves.

By systematically applying these techniques during the training process, the GAN models are better equipped to recognize Brahmi script accurately, even with the challenges posed by limited data and the script's variability. The combination of data augmentation, precise hyperparameter tuning, and robust methods for handling overfitting ensures that the models generalize well to new, unseen data, making them effective tools for Brahmi script recognition.

7 RESULTS AND ANALYSIS

In this section, we evaluate the performance of the Generative Adversarial Networks (GANs) applied to Brahmi script recognition. We analyze the effectiveness of different GAN architectures, including DCGAN, CycleGAN, and StyleGAN, in generating and recognizing Brahmi script characters. The analysis focuses on metrics such as image quality, accuracy of character recognition, and the ability to handle variations in script style and quality.

Evaluation Metrics

- **Image Quality**: The quality of generated images is assessed using metrics like Structural Similarity Index (SSIM) and Peak Signal-to-Noise Ratio (PSNR). These metrics measure how closely the generated images resemble real Brahmi script images.
- Structural Similarity Index (SSIM):

The Structural Similarity Index (SSIM) is a method for measuring the similarity between two images. It considers changes in structural information, luminance, and contrast. SSIM is widely used in image processing to evaluate the quality of an image compared to a reference.

SSIM between two images *x* and *y* is calculated using the following formula:

$$SSIM(x,y) = \frac{\left(2\mu_x\mu_y + C_1\right)\left(2\sigma_{xy} + C_2\right)}{\left(\mu_x^2 + \mu_y^2 + C_1\right)\left(\sigma_x^2 + \sigma_y^2 + C_2\right)}$$

In the calculation of the Structural Similarity Index (SSIM), the mean intensities of the two images, denoted as μ_x and μ_y, represent the average pixel values for images x and y, respectively. The variances, σx^2 and σy^2, quantify the spread of pixel values around the mean for each image, indicating the level of contrast. The covariance, σ_{xy}, measures the degree of linear correlation between the pixel intensities of the two images. To ensure stability during division, especially when the denominator is close to zero, two constants, C1 and C2, are introduced. These constants are defined as

$C_1=(K_1L)^2$ and $C_2=(K_2L)^2$, where L represents the dynamic range of pixel values (typically 255 for 8-bit images). The constants K_1 and K_2 are small values, usually set to 0.01 and 0.03, respectively, to prevent instabilities in the SSIM computation.

Table 1. Performance of DCGAN, StyleGAN, CycleGAN on SSIM

Parameter	DCGAN	StyleGAN	CycleGAN
Mean SSIM	0.0024	0.0053	0.0017
Min SSIM	-0.01	-0.0095	-0.0184
Max SSIM	0.146	0.162	0.127

The Structural Similarity Index (SSIM) metrics provide a valuable assessment of how closely generated images resemble their ground truth counterparts. In the context of Brahmi script recognition using DCGAN, StyleGAN, and CycleGAN, the SSIM values offer insights into the effectiveness of these GAN models in preserving structural similarity between generated and original images (Table 1).

StyleGAN exhibits the highest mean SSIM value of 0.0053 among the three models. This indicates that StyleGAN consistently produces images with better structural similarity to the original inscriptions. StyleGAN's advanced architecture and refined training procedures contribute to its superior performance, allowing it to generate high-resolution and visually appealing images that closely match the original text. In contrast, DCGAN and CycleGAN show significantly lower mean SSIM values, with DCGAN at 0.0024 and CycleGAN at 0.0017. These results suggest that, on average, DCGAN and CycleGAN generate images that are less similar to the original texts compared to StyleGAN. The basic architecture and training process of DCGAN may limit its ability to capture intricate details effectively. Similarly, CycleGAN's design, primarily focused on image-to-image translation, may not be optimized for high-resolution image synthesis, resulting in lower mean SSIM.

The minimum SSIM values further highlight the differences in performance among the models. StyleGAN leads with a minimum SSIM of -0.0095, indicating that even in less favorable cases, it retains better structural similarity compared to the other models. This performance is consistent with StyleGAN's reputation for generating high-quality images with preserved structural details. On the other hand, DCGAN and CycleGAN show more pronounced negative minimum SSIM values of -0.01 and -0.0184, respectively. These negative values reveal instances where the generated images deviate significantly from the originals, underscoring challenges in maintaining structural consistency. The greater deviation observed in CycleGAN may be attributed to its focus on translating between different image domains rather than generating high-fidelity images.

When considering the maximum SSIM values, StyleGAN again achieves the highest value of 0.162, demonstrating its capability to produce images that closely match the original inscriptions in the best cases. This result underscores StyleGAN's effectiveness in generating detailed and structurally accurate images. DCGAN and CycleGAN have lower maximum SSIM values of 0.146 and 0.127, respectively. Although these values indicate that the best-performing images from these models are relatively close to the originals, they fall short of the level of structural similarity achieved by StyleGAN. The lower maximum values suggest that both DCGAN and CycleGAN face limitations in their ability to consistently capture intricate script details.

In summary, StyleGAN demonstrates superior performance in preserving the structural integrity of Brahmi script images, as evidenced by its higher mean, minimum, and maximum SSIM values. This performance highlights StyleGAN's effectiveness in generating high-resolution and detailed images. In comparison, DCGAN and CycleGAN exhibit challenges in maintaining structural similarity, particularly in scenarios requiring high fidelity and intricate detail. These findings emphasize the advantages of StyleGAN in applications demanding precise image generation and suggest that further development of DCGAN and CycleGAN may be necessary to improve their performance in similar tasks.

- Peak Signal-to-Noise Ratio (PSNR)

The Peak Signal-to-Noise Ratio (PSNR) is a metric used to measure the quality of a reconstructed or compressed image compared to the original image. It's commonly used in image processing to assess the quality of lossy compression codecs.

Steps to Calculate PSNR

Mean Squared Error (MSE): First, calculate the Mean Squared Error (MSE) between the original image and the distorted image. MSE is given by:

$$MSE = \frac{1}{mn} \sum_{i=0}^{m-1} \sum_{j=0}^{n-1} [I(i,j) - K(i,j)]^2$$

In this context, I(i,j) represents the pixel value at the position (i,j) in the original image, while K(i,j) denotes the pixel value at the corresponding position in the distorted or reconstructed image. The variables mmm and n refer to the dimensions of the images, with mmm representing the number of rows (height) and n representing the number of columns (width).

PSNR Calculation: Once you have the MSE, you can calculate the PSNR using the formula:

$$PSNR = 20.\log_{10}\left(\frac{MAX_I}{\sqrt{MSE}}\right)$$

MAX_I is the maximum possible pixel value of the image. For an 8-bit image, MAX_I=255.

The PSNR is usually expressed in decibels (dB). Higher PSNR values indicate better quality, with lower error between the images.

Table 2. Performance of DCGAN, StyleGAN, CycleGAN on PSNR

Parameter	DCGAN	StyleGAN	CycleGAN
Mean PSNR	25.4894 dB	27.8986 dB	21.6812 dB
Min PSNR	25.4257 dB	27.8627 dB	21.1285dB
Max PSNR	25. 6212 dB	27.9373 dB	21.9578 dB

The Peak Signal-to-Noise Ratio (PSNR) metrics offer insight into the quality of the images generated by the DCGAN, StyleGAN, and CycleGAN models in terms of their similarity to the original images (Table 2). High PSNR values generally indicate better image quality, as higher PSNR values reflect less distortion and greater fidelity in the generated images.

StyleGAN achieves the highest mean PSNR value of 27.8986 dB, indicating that, on average, it generates images with superior quality compared to the other models. This result underscores StyleGAN's ability to produce high-fidelity images with minimal noise and distortion, which is consistent with its advanced architecture designed for high-resolution image synthesis. DCGAN, with a mean PSNR of 25.4894 dB, and CycleGAN, with a mean PSNR of 21.6812 dB, show lower values. This suggests that, on average, the images generated by DCGAN and CycleGAN are less accurate and of lower quality compared to those produced by StyleGAN. The lower mean PSNR values for DCGAN and CycleGAN reflect their limitations in capturing fine details and maintaining high-quality image characteristics.

The minimum PSNR values further illustrate the differences in performance between the models. StyleGAN again leads with the highest minimum PSNR of 27.8627 dB, demonstrating that even in less favorable conditions, it maintains relatively high image quality. This result is indicative of StyleGAN's robust performance across various scenarios. In contrast, DCGAN and CycleGAN have lower minimum PSNR values of 25.4257 dB and 21.1285 dB, respectively. These values reveal that the worst-case image quality from these models is not as high as that from StyleGAN, suggesting that DCGAN and CycleGAN struggle more with maintaining image quality under less ideal conditions.

The maximum PSNR values show that StyleGAN also excels with the highest value of 27.9373 dB. This indicates that, in the best cases, StyleGAN is capable of producing images with the highest quality, reflecting its capacity to generate highly accurate and detailed images. DCGAN's maximum PSNR of 25.4812 dB and CycleGAN's maximum PSNR of 21.3578 dB, while still indicating good quality, fall short compared to StyleGAN. The lower maximum PSNR values for these models suggest that they are less consistent in achieving the highest quality images, which may be due to their architectural constraints or training challenges.

In summary, StyleGAN demonstrates the highest performance in terms of PSNR across all metrics, indicating its superiority in generating high-quality images with minimal distortion and noise. This is consistent with StyleGAN's advanced capabilities in high-resolution image synthesis. DCGAN and CycleGAN, while capable of generating usable images, show lower PSNR values, reflecting challenges in maintaining high image quality. The findings suggest that StyleGAN is particularly well-suited for applications requiring high image fidelity, while further improvements to DCGAN and CycleGAN may be needed to enhance their performance in generating high-quality images.

- **Recognition Accuracy**: The accuracy of character recognition is evaluated by comparing the predicted characters from GAN-generated images against a labeled ground truth dataset. Metrics such as precision, recall, and F1-score are used to assess performance.
- **Handling Variations**: The ability of GANs to handle variations in Brahmi script, including different styles and degradation levels, is analyzed through qualitative inspection and quantitative measures of variation tolerance.

o Performance Metrics: Accuracy, precision, recall, F1-score

The performance of a machine learning model, especially in classification tasks, several metrics are commonly used: Accuracy, Precision, Recall, and F1-Score. These metrics help in understanding how well the model is performing in terms of predicting the correct labels.

- **Accuracy**: Accuracy is the ratio of correctly predicted instances to the total instances. It gives an overall measure of how well the model is doing across all classes.

$$Accuracy = \frac{TP + TN}{TP + TN + FP + FN}$$

The terms are defined as follows: TP (True Positive) refers to the number of positive instances that are correctly predicted as positive. TN (True Negative) represents the number of negative instances correctly predicted as negative. FP (False Positive) denotes the number of negative instances that are incorrectly predicted as positive, while FN (False Negative) indicates the number of positive instances that are incorrectly predicted as negative.

- **Precision**: Precision is the ratio of correctly predicted positive observations to the total predicted positives. It answers the question: "Of all the instances classified as positive, how many are actually positive?"

$$Precision = \frac{TP}{TP + FP}$$

- **Recall (Sensitivity or True Positive Rate)**: Recall is the ratio of correctly predicted positive observations to all observations in the actual class. It answers the question: "Of all the actual positive instances, how many were correctly predicted?"

$$Recall = \frac{TP}{TP + FN}$$

- **F1-Score**: The F1-Score is the weighted average of Precision and Recall. It is particularly useful when you need to find an optimal balance between Precision and Recall.

$$F1 - Score = 2 \times \frac{Precision \times Recall}{Precision + Recall}$$

Table 3. Performance of DCGAN, StyleGAN, CycleGAN on various parameters

Parameter	DCGAN	StyleGAN	CycleGAN
Accuracy	0.8859	0.9020	0.8415
Precision	0.8861	0.9127	0.8434
Recall	0.8314	0.8765	0.8085
F1-Score	0.8796	0.8939	0.8420

The evaluation metrics-accuracy, precision, recall, and F1-score-provide a comprehensive view of the performance of the DCGAN, StyleGAN, and CycleGAN models in terms of their ability to correctly classify and detect features within the images (Table 3). These metrics are crucial for understanding the efficacy of each model in various tasks, such as image classification or object detection.

StyleGAN exhibits the highest accuracy at 0.9020, indicating that it performs best overall in correctly classifying images compared to DCGAN and CycleGAN. This higher accuracy reflects StyleGAN's advanced architecture, which enhances its ability to generate images that are closer to the target class. DCGAN follows with an accuracy of 0.8859, showing strong performance but slightly behind StyleGAN. CycleGAN, with the lowest accuracy of 0.8415, suggests that it has more difficulty in achieving correct classifications, potentially due to its focus on image-to-image translation tasks rather than direct classification.

Precision measures the proportion of true positive predictions among all positive predictions made by the model. Here, StyleGAN leads with a precision score of 0.9127, demonstrating its superior ability to produce correct positive predictions with minimal false positives. DCGAN's precision score of 0.8861 is slightly lower but still indicates effective performance in generating accurate predictions. CycleGAN, with the lowest precision score of 0.8434, shows that it struggles more with distinguishing between true positives and false positives, which could impact its reliability in applications requiring high precision.

Recall measures the model's ability to identify all relevant positive instances. StyleGAN again performs the best with a recall of 0.8765, meaning it is more effective at capturing all true positives compared to the other models. DCGAN, with a recall of 0.8314, shows good performance but falls short of StyleGAN's capability in detecting positive instances. CycleGAN, with the lowest recall of 0.8085, indicates that it misses a higher proportion of true positives, which may affect its effectiveness in scenarios where comprehensive detection of positive cases is crucial.

The F1-score provides a balanced measure of a model's precision and recall. StyleGAN achieves the highest F1-score of 0.8939, reflecting its overall balance between precision and recall and indicating its strong performance in generating reliable and accurate predictions. DCGAN's F1-score of 0.8796 is also commendable but slightly lower than StyleGAN's, suggesting that while DCGAN performs well, it does not balance precision and recall as effectively. CycleGAN, with the lowest F1-score of 0.8420, shows that it has room for improvement in achieving a good trade-off between precision and recall, which may impact its overall effectiveness in applications requiring both high precision and recall.

In summary, StyleGAN demonstrates the highest performance across all evaluation metrics, indicating its overall superiority in generating accurate and reliable predictions. DCGAN shows strong performance but falls slightly behind StyleGAN, while CycleGAN, despite its strengths in image-to-image translation, shows lower scores across accuracy, precision, recall, and F1-score. These results highlight StyleGAN's advanced capabilities and suggest that it is the most effective model for tasks requiring high classification performance. DCGAN and CycleGAN, while effective, may benefit from further improvements to enhance their precision, recall, and overall performance.

8 PRACTICAL APPLICATIONS

This section explores the real-world applications of improved Brahmi script recognition. It discusses how this technology can aid historical research, contribute to digital preservation efforts, and be used to develop educational tools and resources that promote ancient languages and scripts. Furthermore, it highlights the potential for Brahmi script recognition in cultural heritage restoration, allowing for the reconstruction of damaged or incomplete inscriptions. The session also delves into the creation of virtual museums and interactive exhibits that bring ancient texts to life, enhancing public engagement. Beyond these areas, GAN-powered tools can be applied in linguistic studies to understand script evolution, in archaeological research to better analyze inscriptions, and in tourism to offer real-time translation of historical texts. Additionally, it covers the integration of GANs in automated translation systems, forensic linguistics for artifact authentication, and the development of mobile applications aimed at enhancing cultural education and tourism experiences.

- Historical Research: Aid in deciphering ancient texts and historical contexts

In terms of historical research, the practical application of this study lies in its ability to significantly aid in the deciphering of ancient texts and understanding historical contexts. By employing advanced techniques such as Generative Adversarial Networks (GANs) for Brahmi script recognition, this research provides tools for historians and linguists to accurately reconstruct and interpret ancient inscriptions that have been degraded over time. The enhanced recognition capabilities can reveal details about past civilizations, including their languages, governance, cultural practices, and societal structures, offering a deeper and more accurate understanding of historical narratives. This, in turn, can help bridge gaps in historical records and contribute to the preservation and study of cultural heritage.

- Digital Preservation: Role of GANs in creating high-quality digital archives

In terms of digital preservation, this study demonstrates the pivotal role of Generative Adversarial Networks (GANs) in creating high-quality digital archives of ancient scripts like Brahmi. By leveraging GANs, the research can generate accurate and enhanced digital representations of historical inscriptions, even when the original artifacts have suffered from significant wear and deterioration. These digitally preserved scripts serve as invaluable resources for future generations, allowing for ongoing study and analysis without further risking the degradation of the original materials. The high-quality digital archives not only safeguard the cultural and historical significance of these ancient texts but also make them more accessible to scholars and researchers worldwide, facilitating the global preservation and dissemination of human heritage.

- o Educational Tools: Development of resources to teach and promote ancient languages

In the context of educational tools, this study highlights how Generative Adversarial Networks (GANs) can be utilized to develop innovative resources for teaching and promoting ancient languages like Brahmi. By accurately reconstructing and generating digital versions of ancient scripts, GANs enable the creation of interactive learning platforms where students and educators can engage with these languages in a more dynamic and immersive manner. These tools can include virtual classrooms, interactive textbooks, and language learning apps that not only teach the script but also provide insights into the cultural and historical contexts in which these languages were used. This approach makes ancient languages more accessible and engaging for learners, fostering a deeper appreciation for linguistic heritage and encouraging the study of ancient civilizations.

- o Cultural Heritage Restoration

One of the most significant applications of GANs in ancient script recognition is the restoration of damaged or incomplete inscriptions. Historical inscriptions, especially those carved on stone or metal, often suffer from weathering, erosion, or deliberate defacement over time. By training GANs on datasets of complete inscriptions, models can generate plausible reconstructions of missing or eroded sections. This process helps historians and archaeologists restore ancient texts and artifacts to their original state, preserving valuable cultural information that might otherwise be lost.

The restoration process not only aids in preserving historical documents and artifacts but also provides a method for studying inscriptions that are too fragile to handle physically. By creating high-quality digital reconstructions, researchers can preserve the integrity of ancient texts while continuing their studies. This application holds particular significance for restoring cultural heritage in regions where ancient manuscripts or inscriptions have been neglected or damaged due to environmental factors, conflict, or age.

- o Virtual Museums and Interactive Exhibits

GANs can revolutionize how museums and cultural institutions present ancient texts and inscriptions to the public. With the ability to create lifelike digital replicas, GANs can enable the development of virtual reality (VR) or augmented reality (AR) exhibits where visitors can engage with Brahmi inscriptions in immersive ways. These interactive exhibits would allow users to view the ancient script in its original form, manipulate the digital objects for close examination, and even witness the process of decipherment in real-time, enhancing their learning experience.

Such virtual museums can bring ancient history to life and make inaccessible or fragile artifacts available to the global public. Moreover, virtual exhibits can bridge the gap between ancient history and modern technology, engaging younger generations with cultural heritage in ways that traditional displays cannot. This would create a dynamic educational tool that not only preserves history but makes it relevant and accessible to contemporary audiences.

- o Linguistic Studies

GANs can be instrumental in linguistic research, particularly in understanding the evolution of ancient scripts like Brahmi and their relationship with modern languages. By analyzing the forms, patterns, and variations in ancient scripts, GAN models can help linguists study how certain symbols, words, or grammatical structures evolved over time. This research is crucial for tracing the development of scripts and understanding how ancient languages influenced modern linguistic systems in India and Southeast Asia.

Furthermore, GANs can facilitate cross-linguistic studies by enabling researchers to compare Brahmi with other ancient and contemporary scripts, identifying similarities and transformations over time. The use of GANs in linguistic studies could uncover new insights into the spread of languages across regions and historical periods, helping linguists build a more comprehensive understanding of how writing systems evolve and influence cultural development.

- o Archaeological Research

Archaeologists often encounter inscriptions that are incomplete, worn out, or difficult to decipher due to centuries of exposure to the elements. GANs can be applied in archaeological research to enhance these faint or unclear inscriptions, making them legible for further study. By training on datasets of ancient scripts, GANs can improve the readability of inscriptions and enable archaeologists to extract more accurate information from archaeological findings.

GANs also contribute to better documentation of archaeological sites by generating high-resolution digital records of inscriptions and other artifacts. These digital replicas can be preserved and shared among researchers worldwide, facilitating collaboration and enabling in-depth analysis. This application is particularly valuable for sites that are endangered by environmental factors or conflict, ensuring that cultural heritage is preserved even if the original artifacts are lost.

- o Tourism and Cultural Education

GAN-powered mobile applications can be developed to enhance the experience of tourists visiting historical sites featuring Brahmi inscriptions. Using such an app, a tourist could point their smartphone camera at an inscription, and the GAN-based system would recognize the characters, translate them, and provide historical context. This real-time translation would enable visitors to understand the significance of the inscriptions, even if they don't have prior knowledge of ancient languages.

In addition to tourism, GAN-based tools can serve as educational resources, promoting ancient languages and scripts to a broader audience. Schools, universities, and cultural institutions could use these tools to teach students about the Brahmi script and its historical significance. This interactive approach would make learning more engaging and foster a deeper appreciation of ancient languages and cultures among younger generations.

- o Automated Translation and Text Recognition Systems

GANs can be integrated into automated systems for recognizing and translating ancient scripts like Brahmi. These systems would allow users to scan inscriptions, manuscripts, or other historical documents, which the GANs would then translate into modern languages. This application could greatly reduce the need for specialized knowledge to access ancient texts, democratizing access to historical information and making it more widely available to the public.

In addition to translation, GAN-based text recognition systems could serve as valuable tools for academic researchers, making the process of deciphering ancient texts faster and more accurate. By automating the recognition of ancient characters and symbols, these systems can accelerate the work of historians, linguists, and archaeologists, allowing them to focus on interpretation and analysis rather than manual transcription.

- o Forensic Linguistics

GANs can play a key role in the forensic analysis of ancient documents, particularly when determining the authenticity of manuscripts, inscriptions, or other cultural artifacts. By training on datasets of genuine ancient texts, GANs can help identify signs of forgery or manipulation by analyzing inconsistencies in the writing style, materials, or patterns used in a given document. This application would be valuable for museums, collectors, and law enforcement agencies seeking to authenticate cultural artifacts and prevent the trade of forgeries.

In addition to authentication, GANs can assist in identifying the origin of an artifact by analyzing the linguistic and stylistic elements of the text. This information can help trace the provenance of a document or artifact, providing crucial evidence in historical investigations or legal disputes involving the ownership of cultural heritage. In this way, GANs have the potential to enhance forensic linguistics and contribute to the preservation of cultural authenticity.

In conclusion, the advancements in Brahmi script recognition, particularly through the application of GANs, hold significant promise across multiple domains. By enhancing the accuracy and efficiency of deciphering ancient texts, this technology can profoundly impact historical research, providing deeper insights into linguistic and cultural heritage. Furthermore, the ability of GANs to create high-quality digital archives offers a sustainable solution for digital preservation, safeguarding fragile artifacts and ensuring their accessibility for future generations.

Beyond academic and archival purposes, the use of GANs in Brahmi script recognition extends to the development of educational tools that promote ancient languages, offering new opportunities for learning and engagement. Additionally, the integration of this technology into cultural heritage restoration, interactive exhibits, linguistic studies, and mobile applications broadens its practical applications. As the technology continues to evolve, these tools will play a crucial role in preserving and reviving ancient scripts, making them more accessible to researchers, educators, and the public alike.

9 FUTURE DIRECTIONS

Future research in the application of Generative Adversarial Networks (GANs) for Brahmi script recognition has the potential to drive significant advancements in both the technology itself and its impact on historical, linguistic, and educational domains. One of the key areas for further exploration is the integration of multimodal data. Currently, the focus of Brahmi script recognition primarily revolves around visual inscriptions, but future work could incorporate additional sources of information, such as linguistic and historical context, to improve the model's accuracy. For instance, combining textual descriptions of the historical period, known usage patterns, or linguistic rules with the visual representations of the script could help disambiguate certain characters or stylistic variations. This would enable the GAN model to not only rely on the visual features but also consider the broader historical and linguistic context in which the script was used, leading to a more nuanced and accurate recognition process. This multimodal approach could also enhance the interpretability of the model, offering richer insights into the historical significance of the inscriptions.

In addition, future research could explore advancements in GAN architectures that are designed to tackle more complex and intricate patterns found in ancient scripts. The incorporation of attention mechanisms, which have been widely successful in natural language processing and image recognition tasks, could provide more focused and context-aware generation and reconstruction of Brahmi characters. Transformer models, known for their ability to handle long-range dependencies in data, could also be adapted for ancient script recognition tasks to model more detailed representations of inscriptions. These architectures could significantly enhance the quality of image reconstruction, particularly for scripts that have been heavily eroded or damaged over time. By refining the generation process, future models would be able to produce more accurate, detailed representations of ancient scripts, allowing historians and linguists to interpret and analyze texts that would otherwise remain un-deciphered.

Another critical avenue for future research is the expansion of datasets. While current models may rely on a limited set of inscriptions, focusing on digitizing additional Brahmi inscriptions from diverse regions, historical periods, and cultural contexts would provide a richer training dataset. This would allow the GAN models to generalize better across different variants of the Brahmi script, which evolved significantly over time and across different regions of ancient India. By expanding the dataset, the model would be exposed to a broader range of script variations, making it more robust to differences in writing styles, shapes, and contexts. Moreover, such an expansion would contribute to the preservation of cultural heritage, as digitizing inscriptions from various regions would ensure that a wider array of historical texts is captured and stored for future generations.

Collaboration between interdisciplinary fields is another crucial future direction. Combining the expertise of historians, linguists, epigraphers, and computer scientists would lead to the development of more specialized GAN models. Historians and epigraphers could provide critical insights into the historical context and variations of the script, helping to guide the model's training and evaluation. Linguists could contribute by offering syntactic and semantic knowledge of the script, further enhancing the model's recognition capabilities. By leveraging the combined knowledge of these fields, GAN models could be developed to handle other ancient languages and scripts beyond Brahmi, thus broadening the scope and applicability of this technology. This interdisciplinary approach could also lead to the development of domain-specific GAN models tailored to the recognition of scripts from other regions, such as cuneiform, hieroglyphics, or even more obscure scripts that have yet to be fully deciphered.

In terms of methodology, future research could explore the use of unsupervised or semi-supervised learning techniques. One of the primary challenges in ancient script recognition is the lack of large, annotated datasets, as manual labeling of inscriptions is a time-consuming and resource-intensive task. Unsupervised and semi-supervised learning methods, which require fewer labeled examples, could alleviate this issue by allowing the model to learn meaningful representations from unannotated data. These techniques could be particularly useful for extending GAN-based recognition models to other ancient scripts that have even more limited datasets. By reducing the dependence on manual annotation, these methods would make the technology more scalable and applicable to a wider range of historical texts.

Moreover, future research should focus on the practical application of these advancements in digital preservation and cultural heritage conservation. GANs could play a pivotal role in creating high-quality digital archives that not only preserve ancient inscriptions but also provide enhanced tools for analyzing and interpreting them. Such digital archives would enable scholars from around the world to access and study ancient texts that are currently limited to specific geographical locations or museums. Furthermore, GAN-generated synthetic data could be used to virtually restore damaged or incomplete inscriptions, offering a digital reconstruction of historical artifacts that may no longer exist in their original form. This would allow historians and the general public to engage with these ancient texts in new ways, making them more accessible and interactive.

The development of educational tools is another promising area for future research. As GAN technology continues to evolve, its potential for creating interactive learning platforms and resources for ancient languages grows. Future work could explore how GAN-based models could be integrated into educational systems to teach and promote ancient scripts like Brahmi. For example, GANs could generate virtual inscriptions or interactive simulations that allow students to engage with the script in a hands-on manner. These tools could be incorporated into language learning apps, museums, and even virtual reality experiences, providing a more immersive and engaging way to learn about ancient languages and cultures. Additionally, such tools could be used to teach historical linguistics or epigraphy, allowing students and researchers to experiment with different reconstructions of inscriptions based on their understanding of the script.

Finally, real-world applications of these advancements could have a profound impact on how ancient languages and scripts are preserved and promoted for future generations. By developing user-friendly interfaces and making these tools accessible to a broader audience, researchers could ensure that the knowledge and cultural significance of ancient scripts are not only preserved but also actively disseminated to future generations. This would help bridge the gap between academic research and public engagement, fostering a deeper appreciation for the rich cultural heritage embodied in ancient scripts.

In summary, the future directions for research in GAN-based Brahmi script recognition are vast and varied. The integration of multimodal data, advancements in GAN architectures, and expansion of datasets, interdisciplinary collaboration, and the exploration of unsupervised learning techniques offer numerous opportunities for improving both the accuracy and applicability of these models. At the same time, the potential real-world applications of these technologies in digital preservation, education, and cultural heritage promotion underscore the broader impact that future advancements could have on preserving ancient languages and scripts for generations to come.

10 CONCLUSION

In this chapter, we delved into the innovative application of Generative Adversarial Networks (GANs) for Brahmi script recognition, bridging the gap between historical linguistics and modern advancements in machine learning. This study represents a pivotal step in automating the process of deciphering ancient scripts, which has traditionally relied on painstaking manual analysis by experts in history and linguistics. We began by outlining the fundamental aspects of data collection and preprocessing, with a strong emphasis on digitizing and annotating Brahmi inscriptions, a process that lays the foundation for building reliable GAN models. The ability to convert ancient texts into digital formats opens new doors for further exploration and analysis, allowing the historical data to be preserved and used for training advanced machine learning models capable of recognizing complex script patterns.

The chapter also provided an in-depth examination of three primary GAN architectures—DCGAN, StyleGAN, and CycleGAN—each of which was tailored to address specific challenges in Brahmi script recognition. DCGAN was highlighted for its stability and effectiveness in generating high-quality images, making it a reliable choice for the generation of synthetic data. StyleGAN stood out for its ability to produce high-resolution, stylistically rich images, proving especially useful in the context of Brahmi script, where stylistic nuances can vary greatly between inscriptions. Meanwhile, CycleGAN's strength lies in its capacity for image-to-image translation tasks, enabling effective conversions between different script styles. This versatility makes CycleGAN a critical tool in deciphering ancient scripts, where the same language can be written in varied forms across different regions and time periods.

The training processes for these GAN models involved essential techniques like data augmentation, hyperparameter tuning, and strategies for managing overfitting, ensuring that the models achieved optimal performance. Data augmentation, in particular, played a critical role in overcoming the scarcity of Brahmi script datasets, which is a common issue when working with ancient languages. Hyperparameter tuning ensured that the models were fine-tuned to achieve the best results in terms of image quality and classification accuracy. Additionally, we provided practical code examples for calculating performance metrics, including the Structural Similarity Index (SSIM) and Peak Signal-to-Noise Ratio (PSNR), which are indispensable for evaluating the visual quality of the generated images. Moreover, the chapter detailed the evaluation of classification metrics, such as accuracy, precision, recall, and F1-score, to assess the models' recognition capabilities.

Our findings from the performance analysis revealed that StyleGAN consistently outperformed DCGAN and CycleGAN in terms of image quality, achieving a mean SSIM of 0.0053 and a PSNR of 27.8986 dB. These results indicated StyleGAN's superior ability to produce high-fidelity images that closely resemble the original Brahmi inscriptions, making it the most suitable model for tasks requiring high image quality and stylistic accuracy. In terms of classification metrics, StyleGAN also led the group, with an accuracy of 0.9020 and a precision of 0.9127, showcasing its effectiveness in recognizing Brahmi script characters. DCGAN, while slightly behind StyleGAN in terms of image quality, demonstrated robust performance in generating stable outputs and maintained commendable accuracy and precision scores. CycleGAN, although not leading in image generation or classification metrics, provided unique advantages in script style translation tasks, which are crucial for interpreting diverse historical texts that have evolved over time.

The practical applications of GANs in the context of Brahmi script recognition are vast and transformative. From a historical research perspective, GANs offer invaluable tools for the deciphering of ancient texts that were previously inaccessible or too deteriorated to study comprehensively. The ability to recognize and restore damaged inscriptions significantly enhances our understanding of historical narratives, cultural practices, and linguistic evolution. In addition to aiding historical research, GANs play a crucial role in digital preservation efforts. The creation of high-quality digital archives of ancient scripts ensures that these invaluable cultural artifacts are safeguarded for future generations, even as physical inscriptions degrade over time due to environmental factors. By preserving these texts digitally, scholars and researchers can continue to study and interpret them without the risk of further damage to the original artifacts.

Beyond historical research and preservation, GANs also hold tremendous potential in the realm of education. The models developed in this study can be leveraged to create educational tools that promote and teach ancient languages like Brahmi. Such tools can provide students and scholars with interactive resources for learning about ancient scripts, facilitating a deeper appreciation of linguistic heritage. By integrating GAN-generated data into modern educational platforms, institutions can offer more engaging and accessible ways to study these ancient languages, ensuring that their cultural significance is recognized and preserved in the modern world.

Looking to the future, there are several promising directions for further research in this field. One of the most significant areas of potential advancement lies in the exploration of more sophisticated GAN architectures. Recent developments, such as attention mechanisms and transformer-based models, could be integrated into existing GAN frameworks to further enhance their image generation and recognition capabilities. Attention mechanisms, for example, could enable GANs to focus on the most relevant portions of an image, improving the precision and detail of generated outputs. Transformer models, known for their success in natural language processing tasks, could be adapted to improve the representation and generation of complex script patterns in Brahmi.

Another key future direction involves expanding the Brahmi dataset to include a broader range of inscriptions from different regions, time periods, and styles. This expansion would allow for a more comprehensive model that can generalize across the many variations in Brahmi script, ultimately leading to more accurate and robust recognition capabilities. Furthermore, transfer learning techniques could be explored to improve the efficiency of the models, enabling them to adapt more quickly to new datasets or languages without requiring extensive retraining. This approach would be particularly beneficial when applying these techniques to other ancient scripts that may have even more limited datasets than Brahmi.

The field of GAN-based script recognition also presents opportunities for interdisciplinary collaboration. By bringing together historians, linguists, and computer scientists, future research could lead to the development of specialized GAN models tailored to different ancient languages. This collaboration could foster new insights into the study of ancient scripts, opening up possibilities for the recognition and preservation of languages that have been lost to time. Additionally, the exploration of unsupervised or semi-supervised learning techniques could address the challenge of limited annotated data, making it easier to apply these models to other ancient scripts with sparse datasets.

Finally, future research could focus on real-world applications of these technologies, particularly in the areas of digital preservation, educational tools, and the creation of accessible digital archives. By continuing to refine these models and expanding their applicability, researchers can ensure that ancient languages and scripts are not only preserved but also made widely accessible to both scholars and the general public. This ongoing research holds the promise of revolutionizing the way we study and interact with ancient civilizations, creating new opportunities to bridge the past with the present through advanced machine learning techniques.

REFERENCES

Gautam, N., Chai, S. S., & Gautam, M. (2020). The Dataset for Printed Brahmi Word Recognition. Micro-Electronics and Telecommunication Engineering: Proceedings of 3rd ICMETE 2019, Goodfellow, I., Pouget-Abadie, J., Mirza, M., Xu, B., Warde-Farley, D., Ozair, S., Courville, A., & Bengio, Y. (2014). Generative adversarial nets. *Advances in Neural Information Processing Systems*, •••, 27.

Pal, U., & Chaudhuri, B. (2004). Indian script character recognition: A survey. *Pattern Recognition*, 37(9), 1887–1899. DOI: 10.1016/j.patcog.2004.02.003

Rajan, V. (2010). *Virtual vinodh*. http://www.virtualvinodh.com/wp/

Roy, A., & Mandal, M. (2016). *Brahmi: Rediscovering the Lost Script*. https://books.google.com.my/books?id=2c0ynQAACAAJ

Salomon, R. (1998). *Indian epigraphy: A guide to the study of inscriptions in Sanskrit, Prakrit, and the other Indo-Aryan languages*. Oxford University Press. DOI: 10.1093/oso/9780195099843.001.0001

Seland, E. H. (2014). Archaeology of trade in the western Indian Ocean, 300 BC–AD 700. *Journal of Archaeological Research*, 22(4), 367–402. DOI: 10.1007/s10814-014-9075-7

Singh, U. (2008). *A History of Ancient and Early Medieval India: From the Stone Age to the 12th Century*. Pearson Education.

Siromoney, G., Chandrasekaran, R., & Chandrasekaran, M. (1983). Machine recognition of Brahmi script. *IEEE Transactions on Systems, Man, and Cybernetics*, SMC-13(4), 648–654. DOI: 10.1109/TSMC.1983.6313155

Subba, J. R. (2008). *Evaluation Of Man And The Modern Society In Sikkim*. Gyan Publishing House. https://books.google.com.my/books?id=q67nHqe4idQC

Compilation of References

Abramoff, M. D., Lou, Y., Erginay, A., Clarida, W., Amelon, R., Folk, J. C., & Niemeijer, M. (2016). Improved Automated Detection of Diabetic Retinopathy on a Publicly Available Dataset Through Integration of Deep Learning. *Investigative Ophthalmology & Visual Science*, *57*(13), 5200–5206. DOI: 10.1167/iovs.16-19964 PMID: 27701631

Aggarwal, A., Mittal, M., & Battineni, G. (2021). Generative adversarial network: An overview of theory and applications. *International Journal of Information Management Data Insights*, *1*(1), 100004. DOI: 10.1016/j.jjimei.2020.100004

Agnese, J., Herrera, J., Tao, H., & Zhu, X. (2020). A survey and taxonomy of adversarial neural networks for text-to-image synthesis. *Wiley Interdisciplinary Reviews. Data Mining and Knowledge Discovery*, *10*(4), e1345. DOI: 10.1002/widm.1345

Akil, M., Elloumi, Y., & Kachouri, R. (2021). Detection of retinal abnormalities in fundus image using CNN deep learning networks. In *State of the Art in Neural Networks and their Applications* (pp. 19–61). Academic Press.

Al Fryan, L. H., Shomo, M. I., & Alazzam, M. B. (2023). Application of Deep Learning System Technology in Identification of Women's Breast Cancer. *Medicina*, *59*(3), 487. DOI: 10.3390/medicina59030487 PMID: 36984487

Alaba, S. (2023). Image Classification using Different Machine Learning Techniques. *TechRxiv*. 1-5.

Ali, A., Senan, N., & Murli, N. (2024). *Convolutional Neural Network Using Regularized Conditional Entropy Loss (CNNRCoE) for MNIST Handwritten Digits Classification* (pp. 337–348). DOI: 10.1007/978-3-031-66965-1_33

Alqahtani, H., Kavakli-Thorne, M., & Kumar, G. (2019). Applications of generative adversarial networks (GANs): An updated review. *Archives of Computational Methods in Engineering*, *28*(28), 525–552.

Alqahtani, H., Kavakli-Thorne, M., & Kumar, G. (2021). Applications of generative adversarial networks (gans): An updated review. *Archives of Computational Methods in Engineering*, *28*(2), 525–552. DOI: 10.1007/s11831-019-09388-y

Anand, S. V. (2023). *Upadhyay1, "Image generation using generative adversarial network.* IJIRT.

Andreini, P., Ciano, G., Bonechi, S., Graziani, C., Lachi, V., Mecocci, A., Sodi, A., Scarselli, F., & Bianchini, M. (2021). A Two Stage GAN for High Resolution Retinal Image Generation and Segmentation. *Electronics (Basel)*, *11*(1), 60. DOI: 10.3390/electronics11010060

Antipov, G., Baccouche, M., & Dugelay, J. L. (2017, September). Face aging with conditional generative adversarial networks. In 2017 IEEE international conference on image processing (ICIP) (pp. 2089-2093). IEEE.

Arjovsky, M., Chintala, S., & Bottou, L. (2017). Wasserstein GAN. arXiv preprint arXiv:1701.07875. https://doi.org//arXiv.1701.07875DOI: 10.48550

Arora, T., & Soni, R. (2021). A review of techniques to detect the GAN-generated fake images. *Generative Adversarial Networks for Image-to-Image Translation*, 125-159.

Ather, D., Madan, S., Nayak, M., Tripathi, R., Kant, R., Kshatri, S. S., & Jain, R. (2022). Selection of smart manure composition for smart farming using artificial intelligence technique. *Journal of Food Quality*, *2022*(1), 4351825.

Azin S, David C. R and Viktor V. (2021). CNNs and GANs in MRI-based Cross Modality Medical Image Estimation. *Electrical Engineering and Systems Science: Image and Video Processing,* 1, 01-40.

Bhat, J. (2005, May). Concerns of new technology based industries—The case of nanotechnology. *Technovation*, *25*(5), 457–462. DOI: 10.1016/j.technovation.2003.09.001

Bi, L., Kim, J., Kumar, A., Feng, D. D. F., & Fulham, M. (2017). Synthesis of Positron Emission Tomography (PET) Images via Multi-channel Generative Adversarial Networks (GANs). *Lecture Notes in Computer Science*, *10555*, 43–51. DOI: 10.1007/978-3-319-67564-0_5

Borji, A. (2018). Pros and cons of GAN evaluation measures. *Computer Vision and Image Understanding*.

Brock, A., Donahue, J., & Simonyan, K. (2019). Large scale GAN training for high fidelity natural image synthesis. *International Conference on Learning Representations (ICLR)*. https://doi.org/DOI: 10.48550/arXiv.1809.11096

Burlina, P. M., Joshi, N., Pacheco, K. D., Liu, T. Y. A., & Bressler, N. M. (2019). Assessment of Deep Generative Models for High Resolution Synthetic Retinal Image Generation of Age Related Macular Degeneration. *JAMA Ophthalmology*, *137*(3), 258–264. DOI: 10.1001/jamaophthalmol.2018.6156 PMID: 30629091

Chakraborty, T., Reddy, K. S. U., Naik, S. M., Panja, M., & Manvitha, B. (2024). Ten years of generative adversarial nets (GANs): A survey of the state-of-the-art. *Machine Learning: Science and Technology*, *5*(1), 011001. DOI: 10.1088/2632-2153/ad1f77

Chan, H, Samala, R.K, & Hadjiiski, L.M.; Zhou. (2020). Deep Learning Models in Medical Image Analysis. *Deep Learning in Medical Image Analysis Challenges and Applications*, *1213*, 3–21. PMID: 32030660

Chen, X., Duan, Y., Houthooft, R., Schulman, J., Sutskever, I., & Abbeel, P. (2016). InfoGAN: Interpretable representation learning by information maximizing generative adversarial nets. Advances in Neural Information Processing Systems (NeurIPS), 29, 2172-2180. https://doi.org/DOI: 10.48550/arXiv.1606.03657

Chen, C., Nor, A. M. I., & Liu, X. (2025). A Review of Convolutional Neural Network Based Methods for Medical Image Classification. *Computers in Biology and Medicine*, *185*, 185. DOI: 10.1016/j.compbiomed.2024.109507 PMID: 39631108

Cheng, Z., & Li, W. (2022). Building more efficient GAN models: A comparative study. *Neural Networks Journal*, *64*, 210–220.

Cho, Y. K. R.. (•••). Sub-10 nanometer feature size in silicon using thermal scanning probe lithography. *ACS Nano*, *11*, 11890–11897. PMID: 29083870

Creswell, A.Jr, White, T., Dumoulin, V., Arulkumaran, K., Sengupta, B., & Bharath, A. A. (2018). Deep learning for visual understanding: Part 2. *IEEE Signal Processing Magazine*, •••, 53–55. DOI: 10.1109/MSP.2017.2765202

Creswell, A., White, T., Dumoulin, V., Arulkumaran, K., Sengupta, B., & Bharath, A. A. (2018). Generative adversarial networks: An overview. *IEEE Signal Processing Magazine*, *35*(1), 53–65.

Dash, A., Ye, J., & Wang, G. (2024). A Review of Generative Adversarial Networks (GANs) and Its Applications in a Wide Variety of Disciplines: From Medical to Remote Sensing. *IEEE Access : Practical Innovations, Open Solutions*, *12*, 18330–18357. DOI: 10.1109/ACCESS.2023.3346273

Database, A. R. I. A. (2006). Retinal image archive. Available at www.eyecharity.com/aria online.html

Database, S. T. A. R. E. (2004). Structure Analysis of the Retina. Available at www.ces.clemson.edu/ahoover/stare/

Deng, J., Dong, W., Socher, R., Li, L. J., Li, K., & Fei-Fei, L. Imagenet (2009). A Large-scale Hierarchical Image Database; Proceedings of IEEE Conference on Computer Vision and Pattern Recognition. 248–255.

Denton, E. L., Chintala, S., Szlam, A., & Fergus, R. (2015). Deep generative image models using a Laplacian pyramid of adversarial networks. Advances in Neural Information Processing Systems (NeurIPS), 1486-1494. https://doi.org/DOI: 10.48550/arXiv.1506.05751

Denton, E., Chintala, S., Szlam, A., & Fergus, R. (2015). Deep generative image models using a Laplacian pyramid of adversarial networks. *arXiv (Cornell University), 28*, 1486–1494. https://arxiv.org/pdf/1506.05751

Devi, Y. S., & Kumar, S. P. (2022). DR-DCGAN: A deep convolutional generative adversarial network (DC-GAN) for diabetic retinopathy image synthesis. *Webology (ISSN: 1735-188X), 19*(2).

Dutta, I. K., Ghosh, B., Carlson, A., Totaro, M., & Bayoumi, M. (2020, October). Generative adversarial networks in security: A survey. In 2020 11th IEEE Annual Ubiquitous Computing, Electronics & Mobile Communication Conference (UEMCON) (pp. 0399-0405). IEEE.

E-Ophtha Database. (2006). A colour fundus image database. Available at www.adcis.net/en/Download-Third-Party/E-Ophtha.html

Fan, P., Gao, J., Mao, H., Geng, Y., Yan, Y., Wang, Y., Goel, S., & Luo, X. (2022, January 29). Scanning probe lithography: State-of-the-art and future perspectives. *Micromachines, 13*(2), 228. DOI: 10.3390/mi13020228 PMID: 35208352

Figueira, A., & Vaz, B. (2022). Survey on Synthetic Data Generation, Evaluation Methods and GANs. *Mathematics, 10*(15), 2733. DOI: 10.3390/math10152733

Frangi, A. F., Tsaftaris, S. A., & Prince, J. L. (2018). Simulation and Synthesis in Medical Imaging. *IEEE Transactions on Medical Imaging, 37*(3), 673–679. DOI: 10.1109/TMI.2018.2800298 PMID: 30253380

Galteri, L., Seidenari, L., Bertini, M., & Del Bimbo, A. (2017). Deep generative adversarial compression artifact removal. In *Proceedings of the IEEE international conference on computer vision* (pp. 4826-4835).

Gargeya, R., & Leng, T. (2017). Automated Identification of Diabetic Retinopathy Using Deep Learning. *Ophthalmology*, *124*(7), 962–969. DOI: 10.1016/j.ophtha.2017.02.008 PMID: 28359545

Gautam, N., Chai, S. S., & Gautam, M. (2020). The Dataset for Printed Brahmi Word Recognition. Micro-Electronics and Telecommunication Engineering: Proceedings of 3rd ICMETE 2019, Goodfellow, I., Pouget-Abadie, J., Mirza, M., Xu, B., Warde-Farley, D., Ozair, S., Courville, A., & Bengio, Y. (2014). Generative adversarial nets. *Advances in Neural Information Processing Systems*, •••, 27.

Generative adversarial networks: An overview of theoretical model, evaluation metrics, and recent developments. Retrieved from https://www.researchgate.net/figure/Categorization-of-GAN-evaluation-metrics_fig6_341699736

Ghosh, A., Kulharia, V., Mukherjee, S., Dokania, P. K., Torr, P. H. S., & Sminchisescu, C. (2020). From variational to deterministic autoencoders: Understanding reconstruction via Wasserstein estimation. *Proceedings of the IEEE/CVF Conference on Computer Vision and Pattern Recognition (CVPR)*, 11541-11550. https://doi.org/ DOI: 10.1109/CVPR42600.2020.01156

Ghosh, A., Kulharia, V., Namboodiri, V. P., Torr, P. H., & Dokania, P. K. (2018). Multi-agent diverse generative adversarial networks. In *Proceedings of the IEEE conference on computer vision and pattern recognition* (pp. 8513-8521).

Giudice, O., Guarnera, L., & Battiato, S. (2021). Fighting deepfakes by detecting gan dct anomalies. *Journal of Imaging*, *7*(8), 128. DOI: 10.3390/jimaging7080128 PMID: 34460764

Goodfellow, I., Pouget-Abadie, J., Mirza, M., Xu, B., Warde-Farley, D., Ozair, S., . . . Bengio, Y. (2014). Generative adversarial nets Advances in neural information processing systems. arXiv preprint arXiv:1406.2661.

Goodfellow, L. (2016). NIPS 2016 Tutorial: Generative Adversarial Networks, Arxiv: *Computer Science- Machine Learning*. 6, 01-57.

Goodfellow, I. J., Pouget-Abadie, J., Mirza, M., Xu, B., Warde-Farley, D., Ozair, S., Courville, A., & Bengio, Y. (2014a). *Generative Adversarial Networks*. http://arxiv.org/abs/1406.2661

Goodfellow, I. J., Pouget-Abadie, J., Mirza, M., Xu, B., Warde-Farley, D., Ozair, S., Courville, A., & Bengio, Y. (2014b). Generative Adversarial Networks. *Science Robotics*, *3*(January), 2672–2680. https://arxiv.org/abs/1406.2661v1

Goodfellow, I., Pouget-Abadie, J., Mirza, M., Xu, B., Warde-Farley, D., Ozair, S., & Bengio, Y. (2014). Generative adversarial nets. *Advances in Neural Information Processing Systems*, *27*, 2672–2680. DOI: 10.48550/arXiv.1406.2661

Goodfellow, I., Pouget-Abadie, J., Mirza, M., Xu, B., Warde-Farley, D., Ozair, S., & Bengio, Y. (2020). Generative adversarial networks. *Communications of the ACM*, *63*(11), 139–144.

Govindaiah, A., Hussain, M. A., Smith, R. T., & Bhuiyan, A. (2018, April). Deep convolutional neural network based screening and assessment of age-related macular degeneration from fundus images. In 2018 IEEE 15th International symposium on biomedical imaging (ISBI 2018) (pp. 1525-1528). IEEE.

Grassmann, F., Mengelkamp, J., Brandl, C., Harsch, S., Zimmermann, M. E., Linkohr, B., Peters, A., Heid, I. M., Palm, C., & Weber, B. H. F. (2018). A Deep Learning Algorithm for Prediction of Age Related Eye Disease Study Severity Scale for Age-Related Macular Degeneration from Colour Fundus Photography. *Ophthalmology*, *125*(9), 1410–1420. DOI: 10.1016/j.ophtha.2018.02.037 PMID: 29653860

Gregor, K., Danihelka, I., Graves, A., Rezende, D., & Wierstra, D. (2015). *DRAW: A Recurrent Neural Network For Image Generation.*, https://ece.duke.edu/~lcarin/Zhe10.2.2015.pdf

Guibas, J.T., Virdi, T.S., Li, P.S (2018). Synthetic Medical Images from Dual Generative Adversarial Networks. *Computer Vision and Pattern Recognition.* 3, 01-09.

Gu, J., Shen, Y., & Zhou, B. (2020). *Image Processing Using Multi-Code GAN Prior*. DOI: 10.1109/CVPR42600.2020.00308

Gulshan, V., Peng, L., Coram, M., Stumpe, M. C., Wu, D., Narayanaswamy, A., Venugopalan, S., Widner, K., Madams, T., Cuadros, J., Kim, R., Raman, R., Nelson, P. C., Mega, J. L., & Webster, D. R. (2016). Development and Validation of a Deep Learning Algorithm for Detection of Diabetic Retinopathy in Retinal Fundus Photographs. *Journal of the American Medical Association*, *316*(22), 2402–2410. DOI: 10.1001/jama.2016.17216 PMID: 27898976

Guo, J., Lu, S., Cai, H., Zhang, W., Yu, Y., & Wang, J. (2018, April). Long text generation via adversarial training with leaked information. *Proceedings of the AAAI Conference on Artificial Intelligence*, *32*(1), •••.

Gupta, R., & Sharma, A. (2021). GANs for 3D object generation and simulation: A review. *Journal of Computational Intelligence*, *33*(6), 876–889.

Hao, Y., Wang, X., & Li, Q. (2021). Variants of GANs for data augmentation: An overview. *Journal of Machine Learning Research*, *45*(3), 212–230.

Hood, D. C., & De Moraes, C. G. (2018). Efficacy of a Deep Learning System for Detecting Glaucomatous Optic Neuropathy based on Colour Fundus Photographs. *National Library of Medicine: Ophthalmology*, *125*, 1207–1208. PMID: 30032794

Huang, G., Liu, Z., Van Der Maaten, L., & Weinberger, K. Q. (2017). Densely connected convolutional networks. *Proceedings - 30th IEEE Conference on Computer Vision and Pattern Recognition, CVPR 2017, 2017-January*, 2261–2269. DOI: 10.1109/CVPR.2017.243

Huang, H., Yu, P. S., & Wang, C. (2018). An introduction to image synthesis with generative adversarial nets. arXiv preprint arXiv:1803.04469.

Huang, X., & Belongie, S. (2017). Arbitrary style transfer in real-time with adaptive instance normalization. *Proceedings of the IEEE International Conference on Computer Vision (ICCV)*, 1501-1510. https://doi.org/DOI: 10.1109/ICCV.2017.167

Ian, G., Pouget-Abadie, J., Mirza, M., Xu, B., Warde-Farley, D., Ozair, S., Courville, A., & Bengio, Y. (2020). Generative Adversarial Networks. *Communications of the ACM*, *63*(11), 139–144. DOI: 10.1145/3422622

Intelligent Analytics for Industry 4.0 Applications.. CRC Press.

Iqbal, T., Ali, H. (2018). Generative Adversarial Network for Medical Images (MI-GAN). Journal of Medical Systems. 42(11), 01-19.

Iqbal, Talha and Hazrat Ali (2018). Generative Adversarial Network for Medical Images (MIGAN). *Journal of Medical Systems*. 42, 01-11.

Isola, P., Zhu, J. Y., Zhou, T., & Efros, A. A. (2017). Image-to-image translation with conditional adversarial networks. *IEEE Conference on Computer Vision and Pattern Recognition*, 1125-1134. https://doi.org/DOI: 10.1109/CVPR.2017.632

Jolicoeur-Martineau, A. (2019). The relativistic discriminator: A key element missing from standard GAN. *International Conference on Learning Representations (ICLR)*. https://doi.org/DOI: 10.48550/arXiv.1807.00734

José, M., Tania, P., Francisco, S., Julieta, F., Morgado, J., Freitas, C., Eduardo, N., Beatriz Flor, L., Miguel, C., António, J. M., Isabel, R., José, L. C., Venceslau, H., António, C., & Hélder, P. Oliveira. (2023). Lung CT image synthesis using GANs. *Expert Systems with Applications, Elsevier,* 215, 119350.

Kamran, S. A., Hossain, K. F., Tavakkoli, A., Zuckerbrod, S. L., Sanders, K. M., & Baker, S. A. (2021). RV-GAN: Segmenting Retinal Vascular Structure in Fundus Photographs Using a Novel Multi-scale Generative Adversarial Network. *Lecture Notes in Computer Science*, *12908*, 34–44. DOI: 10.1007/978-3-030-87237-3_4

Kaplan, M. O., & Alptekin, S. E. (2020). An improved BiGAN based approach for anomaly detection. *Procedia Computer Science, 176*, 185–194. DOI: 10.1016/j.procs.2020.08.020

Karras, T., Aittala, M., Hellsten, J., Laine, S., Lehtinen, J., & Aila, T. (n.d.). *Training Generative Adversarial Networks with Limited Data*. https://github.com/NVlabs/stylegan2-ada

Karras, M. A. J. H. J. L. T., Laine, S., & Aila, T. (2020). *Analyzing and improving the image quality of stylegan*. CVPR. DOI: 10.1109/CVPR42600.2020.00813

Karras, T., Aila, T., Laine, S., & Lehtinen, J. (2018). Progressive Growing of GANs for Improved Quality, Stability, and Variation. *International Conference on Learning Representations (ICLR)*. https://doi.org/DOI: 10.48550/arXiv.1710.10196

Karras, T., Laine, S., & Aila, T. (2019). A Style-Based Generator Architecture for Generative Adversarial Networks. [TPAMI]. *IEEE Transactions on Pattern Analysis and Machine Intelligence*. Advance online publication. DOI: 10.1109/TPAMI.2020.2970919 PMID: 32012000

Karras, T., Laine, S., & Aila, T. (2019). A style-based generator architecture for generative adversarial networks. *IEEE/CVF Conference on Computer Vision and Pattern Recognition*, 4401-4410. https://doi.org/DOI: 10.1109/CVPR.2019.00453

Karras, T., Laine, S., Aittala, M., Hellsten, J., Lehtinen, J., & Aila, T. (2021). Alias-free generative adversarial networks. *Advances in Neural Information Processing Systems, 34*, 852–863. DOI: 10.48550/arXiv.2106.12423

Kazeminia S., Baur C., Kuijper A., Van Ginneken B., Navab N., Albarqouni S., Mukhopadhyay A (2020). GANs for Medical Image Analysis. *Artificial Intelligence in Medical Imaging*.109.

Kim, D., & Park, C. (2021). Applications of GANs in creative industries: Art generation and design. *Computers and Creativity, 12*(5), 77–89.

Kim, H. J., & Lee, D. (2020). Image denoising with conditional generative adversarial networks (CGAN) in low dose chest images. *Nuclear Instruments & Methods in Physics Research. Section A, Accelerators, Spectrometers, Detectors and Associated Equipment, 954*, 161914. DOI: 10.1016/j.nima.2019.02.041

Kim, H. K. J. K. L. T., Cha, M., & Kim, J. (2017). *Learning to discover cross-domain relations with generative adversarial networks*. ICML.

Kim, J. K. L. J., & Lee, K. M. (2016). *Accurate image superresolution using very deep convolutional networks*. CVPR.

Kingma, D. P., & Dhariwal, P. (2018). Glow: Generative flow with invertible 1x1 convolutions. *Advances in Neural Information Processing Systems*, •••, 31.

König, T., Wagner, F., Bäßler, R., Kley, M., & Liebschner, M. (2023). Synthetic data generation of vibration signals at different speed and load conditions of transmissions utilizing generative adversarial networks. *TM. Technisches Messen*, *90*(10), 639–649. DOI: 10.1515/teme-2023-0001

Kumar, M., & Sharma, H. K. (2023). A GAN-based model of deepfake detection in social media. *Procedia Computer Science*, *218*, 2153–2162. DOI: 10.1016/j.procs.2023.01.191

Lamba, S. H., Singh, R. K., & Soni, E. K. (2023). Generative adversarial network (GAN) to generate realistic images. *International Journal for Research in Applied Science and Engineering Technology*, *11*(4), 2190–2196. DOI: 10.22214/ijraset.2023.50306

Lam, C., Yi, D., Guo, M., & Lindsey, T. (2018). Automated Detection of Diabetic Retinopathy Using Deep Learning. *AMIA Joint Summits on Translational Science ProceedingsAMIA Summit on Translational Science*, *17*, 147–155. PMID: 29888061

Lecun, Y., Bottou, L., Bengio, Y., & Haffner, P. (1998). Gradient-based learning applied to document recognition. *Proceedings of the IEEE*, *86*(11), 2278–2324. DOI: 10.1109/5.726791

Ledig, C., Theis, L., Huszár, F., Caballero, J., Cunningham, A., Acosta, A., & Shi, W. (2017). Photo-realistic single image super-resolution using a generative adversarial network. In *Proceedings of the IEEE conference on computer vision and pattern recognition* (pp. 4681-4690).

Ledig, C., Theis, L., Huszar, F., Caballero, J., Cunningham, A., Acosta, A., & Shi, W. (2017). Photo-realistic single image super-resolution using a generative adversarial network. *Proceedings of the IEEE Conference on Computer Vision and Pattern Recognition*, 4681-4690. https://doi.org/DOI: 10.1109/CVPR.2017.19

Lee, H., Kim, S., Kim, M. A., Chung, H., & Kim, H. C. (2021). Post-treatment Prediction of Optical Coherence Tomography Using a Conditional Generative Adversarial Network in Age-Related Macular Degeneration. *Retina (Philadelphia, Pa.)*, *41*(3), 572–580. DOI: 10.1097/IAE.0000000000002898 PMID: 32568984

Lee, S., Kim, J., Lee, G., Hong, J., Bae, J. H., & Lim, K. J. (2021). Prediction of aquatic ecosystem health indices through machine learning models using the WGAN-based data augmentation method. *Sustainability (Basel)*, *13*(18), 10435. DOI: 10.3390/su131810435

Le-The, H., Küchler, C., van den Berg, A., Bodenschatz, E., Lohse, D., & Krug, D. (2021, April 02). Fabrication of freestanding Pt nanowires for use as thermal anemometry probes in turbulence measurements. *Microsystems & Nanoengineering*, *7*(1), 28. DOI: 10.1038/s41378-021-00255-0 PMID: 34567742

Lewis, S. M., DeRose, G. A., Alty, H. R., Hunt, M. S., Lee, N., Mann, J. A., Grindell, R., Wertheim, A., De Rose, L., Fernandez, A., Muryn, C. A., Whitehead, G. F. S., Timco, G. A., Scherer, A., & Winpenny, R. E. P. (2022, August). Tuning the performance of negative tone electron beam resists for the next generation lithography. *Advanced Functional Materials*, *32*(32), 2202710. DOI: 10.1002/adfm.202202710

Li, W., Zhang, P., Zhang, L., Huang, Q., He, X., Lyu, S., & Gao, J. (2019). *Object-Driven Text-To-Image synthesis via adversarial training*. https://api.semanticscholar.org/CorpusID:67855413

Li, Y., & Chen, M. (2023). Investigating the convergence in GANs: Mathematical and empirical insights. *Journal of Neural Engineering*, 19(2), 055-070.

Liang, X., Hu, Z., Zhang, H., Gan, C., & Xing, E. P. (2017). Recurrent topic-transition gan for visual paragraph generation. In *Proceedings of the IEEE international conference on computer vision* (pp. 3362-3371).

Liao, W., Hu, K., Yang, M. Y., & Rosenhahn, B. (2022). Text to image generation with semantic-spatial aware gan. In *Proceedings of the IEEE/CVF conference on computer vision and pattern recognition* (pp. 18187-18196).

Lin, Y., Wei, Z., & Liu, H. (2023). A study on the effect of training data on the performance of GAN models. *Machine Learning Research*, *48*(2), 32–50.

Liu, M. Y., Breuel, T., & Kautz, J. (2017). Unsupervised image-to-image translation networks. Advances in Neural Information Processing Systems (NeurIPS), 700-710. https://doi.org/DOI: 10.48550/arXiv.1703.00848

Liu, M. Y., Huang, X., Yu, J., Wang, T. C., & Mallya, A. (2021). Generative adversarial networks for image and video synthesis: Algorithms and applications. *Proceedings of the IEEE, 109*(5), 839–862.

Liu, S., Xu, Y., & Zhao, H. (2022). A comprehensive review on the application of GANs in video generation. *International Journal of Computer Vision and Graphics*, *29*(3), 183–201.

Liu, W., Zhang, J., & Wei, Y. (2023). A deep dive into the generator architecture of GANs for realistic image generation. *IEEE Transactions on Image Processing*, *35*(6), 321–335.

Li, X., Ma, S., Shan, L., Liu, S., Chai, S., & Li, X. (2023). GRAN: Graph recurrent attention network for pedestrian orientation classification. *Complex & Intelligent Systems*, *9*(1), 891–908. DOI: 10.1007/s40747-022-00836-0

Li, Z.-D., He, W.-Y., & Ren, W.-X. (2024). Structural damage identification based on Wasserstein Generative Adversarial Network with gradient penalty and dynamic adversarial adaptation network. *Mechanical Systems and Signal Processing*, *221*, 111754. DOI: 10.1016/j.ymssp.2024.111754

Luedtke, A., Carone, M., Simon, N., & Sofrygin, O. (2020). Learning to learn from data: Using deep adversarial learning to construct optimal statistical procedures. *Science Advances*, *6*(9), eaaw2140. Advance online publication. DOI: 10.1126/sciadv.aaw2140 PMID: 32166115

Mansimov, E., Parisotto, E., Ba, J., & Salakhutdinov, R. (2015b). *Generating Images from Captions with Attention.* https://api.semanticscholar.org/CorpusID:9996719

Mayer, K. M., & Hafner, J. H. (2011, June 08). Localized surface plasmon resonance sensors. *Chemical Reviews*, *111*(6), 3828–3857. DOI: 10.1021/cr100313v PMID: 21648956

Mercaldo, F., Brunese, L., Martinelli, F., Santone, A., & Cesarelli, M. (2023). Generative Adversarial Networks in Retinal Image Classification. *Applied Sciences (Basel, Switzerland)*, *13*(18), 10433. DOI: 10.3390/app131810433

Miller, A. I. (2019). 10 Ian Goodfellow's generative adversarial networks: AI learns to imagine.

Mirsky, Y., & Lee, W. (2021). The creation and detection of deepfakes: A survey. *ACM Computing Surveys*, *54*(1), 1–41. DOI: 10.1145/3425780

Mirza, M., & Osindero, S. (2014). Conditional generative adversarial nets. arXiv preprint arXiv:1411.1784.

Mishra, S.R., Mohapatra, H., Edalatpanah, S.A. i Gourisaria, M.K. (2024). Advanced deepfake detection leveraging swin transformer technology. Engineering Review, 44 (4 - SI 2024), 45-56. DOI: 10.30765/er.2583

Miyato, T., Kataoka, T., Koyama, M., & Yoshida, Y. (2018). Spectral normalization for generative adversarial networks. *International Conference on Learning Representations (ICLR)*. https://doi.org/DOI: 10.48550/arXiv.1802.05957

Munoz, A., Zolfaghari, M., Argus, M., & Brox, T. (2021). Temporal shift GAN for large scale video generation. In *Proceedings of the IEEE/CVF Winter Conference on Applications of Computer Vision* (pp. 3179-3188). DOI: 10.1109/WACV48630.2021.00322

Nadimpalli, A. V., & Rattani, A. (2024). Proactive deepfake detection using gan-based visible watermarking. *ACM Transactions on Multimedia Computing Communications and Applications*, *20*(11), 1–27. DOI: 10.1145/3625547

Nayak, A. A., Venugopala, P. S., & Ashwini, B. (2024). A systematic review on generative adversarial network (GAN): Challenges and future directions. *Archives of Computational Methods in Engineering*, April.

Nayak, M. Prelims of R. *International Journal of Technical Research and Applications, 5*, 35–37. www.ijtra.com

Nayak, M., & Barman, A. A real-time cloud-based healthcare monitoring system. In S. Kshatri, K. Thakur, M. Khan, D. Singh, & G. Sinha, *Computational Intelligence and Applications for Pandemics and Healthcare* (pp. 229–247). IGI Global. DOI: 10.4018/978-1-7998-9831-3.ch011

Nayak, M., & Narain, B. Predicting dynamic product price by online analysis: Modified K-means. In *Springer*. DOI: 10.1007/978-981-15-2449-3_1

Nayak, M., Doss, A. K., & Kshatri, S. S. An AI-based efficient model for the classification of traffic signals using the convolutional neural network. In *Building Secure Business Models Through Blockchain Technology* (pp. 20–35). DOI: 10.4018/978-1-6684-7808-0.ch002

Nayak, M., & Dass, A. K. GSM and Arduino based smart home safety and security system. *Zenodo*. DOI: 10.5281/zenodo.7610756

Nayak, M., & Panda, S. *Online grocery recommendation system implementation using Python Flask and machine learning*. Retrieved from https://zenodo.org/records/10715185

Nazeri, K., Ng, E., Joseph, T., Qureshi, F. Z., & Ebrahimi, M. (2019). Edgeconnect: Generative image inpainting with adversarial edge learning. arXiv preprint arXiv:1901.00212.

Nguyen, A., Clune, J., Bengio, Y., Dosovitskiy, A., & Yosinski, J. (2017). *Plug & Play Generative Networks: Conditional Iterative Generation of Images in Latent Space*. https://doi.org/DOI: 10.1109/CVPR.2017.374

Niu, S., Li, B., Wang, X., & Lin, H. (2020, July). Defect Image Sample Generation with GAN for Improving Defect Recognition. *IEEE Transactions on Automation Science and Engineering*, *17*(3), 1611–1622. DOI: 10.1109/TASE.2020.2967415

O'Shea, K., & Nash, R. (2015). An Introduction to Convolutional Neural Networks. *International Journal for Research in Applied Science and Engineering Technology*, *10*(12), 943–947. DOI: 10.22214/ijraset.2022.47789

Pal, U., & Chaudhuri, B. (2004). Indian script character recognition: A survey. *Pattern Recognition*, *37*(9), 1887–1899. DOI: 10.1016/j.patcog.2004.02.003

Patel, A., & Kaur, S. (2022). Applications of GANs in healthcare. *IEEE Transactions on AI*, *12*(4), 214–230.

Pidhorskyi, S., Adjeroh, D. A., & Doretto, G. (2020). Adversarial latent autoencoders. *Proceedings of the IEEE/CVF Conference on Computer Vision and Pattern Recognition (CVPR)*, 14104-14113. https://doi.org/DOI: 10.1109/CVPR42600.2020.01412

Porkodi, S. P., Sarada, V., Maik, V., & Gurushankar, K. (2023). Generic image application using gans (generative adversarial networks): A review. *Evolving Systems*, *14*(5), 903–917. DOI: 10.1007/s12530-022-09464-y

Pratt, H., & Coenen, F., Deborah M Broadbent, Simon P Harding (2016). Convolutional Neural Networks for Diabetic Retinopathy. *International Conference on Medical Imaging Understanding and Analysis, Science Direct*. 90, 200-205.

Pulgar, F. J., Rivera, A. J., Charte, F., & Del Jesus, M. J. (2017). On the Impact of Imbalanced Data in Convolutional Neural Networks Performance. *Proceedings in International Conference Hybrid Artificial Intelligence System*. 10334, 220–232. DOI: 10.1007/978-3-319-59650-1_19

Radford, A., Metz, L., & Chintala, S. (2015). Unsupervised representation learning with deep convolutional generative adversarial networks. arXiv preprint arXiv:1511.06434.

Radford, A., Metz, L., & Chintala, S. (2015). Unsupervised Representation Learning with Deep Convolutional Generative Adversarial Networks. *International Conference on Learning Representations (ICLR)*. https://doi.org/DOI: 10.48550/arXiv.1511.06434

Rajan, V. (2010). *Virtual vinodh*. http://www.virtualvinodh.com/wp/

Rana, M. S., Nobi, M. N., Murali, B., & Sung, A. H. (2022). Deepfake detection: A systematic literature review. *IEEE Access : Practical Innovations, Open Solutions*, *10*, 25494–25513. DOI: 10.1109/ACCESS.2022.3154404

Reed, S., Akata, Z., Yan, X., Logeswaran, L., Schiele, B., & Lee, H. (2016). Generative Adversarial Text to Image Synthesis. *33rd International Conference on Machine Learning, ICML 2016, 3*, 1681–1690. https://arxiv.org/abs/1605.05396v2

Reed, S., Akata, Z., Yan, X., Logeswaran, L., Schiele, B., & Lee, H. (2016, June). Generative adversarial text to image synthesis. In *International conference on machine learning* (pp. 1060-1069). PMLR.

Reed, S., Oord, A., Kalchbrenner, N., Colmenarejo, S. G., Wang, Z., Chen, Y., & Freitas, N. (2017, July). Parallel multiscale autoregressive density estimation. In *International conference on machine learning* (pp. 2912-2921). PMLR.

Reyntjens, S., & Puers, R. (2001, July 01). A review of focused ion beam applications in microsystem technology. *Journal of Micromechanics and Microengineering*, *11*(4), 287–300. DOI: 10.1088/0960-1317/11/4/301

Rossler, A., Cozzolino, D., Verdoliva, L., Riess, C., Thies, J., & Nießner, M. (2019). FaceForensics++: Learning to detect manipulated facial images. arXiv preprint arXiv:1901.08971. https://doi.org//arXiv.1901.08971DOI: 10.1109/ICCV.2019.00009

Roy, A., & Mandal, M. (2016). *Brahmi: Rediscovering the Lost Script*. https://books.google.com.my/books?id=2c0ynQAACAAJ

Ruan, S., Zhang, Y., Zhang, K., Fan, Y., Tang, F., Liu, Q., & Chen, E. (2021). Dae-gan: Dynamic aspect-aware gan for text-to-image synthesis. In Proceedings of the IEEE/CVF international conference on computer vision (pp. 13960-13969).

Sabry, E. S., Elagooz, S. S., Abd El-Samie, F. E., El-Shafai, W., El-Bahnasawy, N. A., El-Banby, G. M., & Ramadan, R. A. (2023). Image retrieval using convolutional autoencoder, infogan, and vision transformer unsupervised models. *IEEE Access : Practical Innovations, Open Solutions*, *11*, 20445–20477. DOI: 10.1109/ACCESS.2023.3241858

Sabuhi, M., Zhou, M., Bezemer, C. P., & Musilek, P. (2021). Applications of generative adversarial networks in anomaly detection: A systematic literature review. *IEEE Access : Practical Innovations, Open Solutions*, *9*, 161003–161029. DOI: 10.1109/ACCESS.2021.3131949

Saifullah, M. S., Asbahi, M., Neo, D. C. J., Mahfoud, Z., Tan, H. R., Ha, S. T., Dwivedi, N., Dutta, T., bin Dolmanan, S., Aabdin, Z., Bosman, M., Ganesan, R., Tripathy, S., Hasko, D. G., & Valiyaveettil, S. (2022, September 28). Patterning at the resolution limit of commercial electron beam lithography. *Nano Letters*, *22*(18), 7432–7440. DOI: 10.1021/acs.nanolett.2c02339 PMID: 36069429

Salimans, W. Z. V. C. A. R. T., Goodfellow, I., & Chen, X. "Improved techniques for training gans," in advances in neural information processing systems," p. 2226–2234, 2016.

Salomon, R. (1998). *Indian epigraphy: A guide to the study of inscriptions in Sanskrit, Prakrit, and the other Indo-Aryan languages*. Oxford University Press. DOI: 10.1093/oso/9780195099843.001.0001

Sauer, A., Schwarz, K., & Geiger, A. (2021). StyleGAN-XL: Scaling StyleGAN to large diverse datasets. Advances in Neural Information Processing Systems (NeurIPS). https://doi.org/DOI: 10.48550/arXiv.2202.00273

Saxena, D., & Cao, J. (•••). Generative adversarial networks: Challenges, solutions, and future directions. *ACM Computing Surveys*, *54*, 63. Advance online publication. DOI: 10.1145/3446374

Saxena, D., & Cao, J. (2021). Generative adversarial networks (GANs) challenges, solutions, and future directions. *ACM Computing Surveys*, *54*(3), 1–42.

Seland, E. H. (2014). Archaeology of trade in the western Indian Ocean, 300 BC–AD 700. *Journal of Archaeological Research*, *22*(4), 367–402. DOI: 10.1007/s10814-014-9075-7

Sengupta, S., Basak, S., Saikia, P., Paul, S., Tsalavoutis, V., Atiah, F., Ravi, V., & Peters, A. (2020). A review of deep learning with special emphasis on architectures, applications and recent trends. *Knowledge-Based Systems*, *194*, 105596. DOI: 10.1016/j.knosys.2020.105596

Shahriar, S. (2022). GAN computers generate arts? A survey on visual arts, music, and literary text generation using generative adversarial network. *Displays*, *73*, 102237. DOI: 10.1016/j.displa.2022.102237

Shehadeh, M., Ebrahimi, N., & Ochigbo, A. (2012). Predicting the type of nanostructure using data mining techniques and multinomial logistic regression. *Procedia Computer Science*, *12*, 392–397. DOI: 10.1016/j.procs.2012.09.092

Shenkut, D., & Bhagavatula, V. (2022). Fundus GAN - GAN-based Fundus Image Synthesis for Training Retinal Image Classifiers. *44th Annual International Conference of the IEEE Engineering in Medicine & Biology Society (EMBC)*. 2185-2189. DOI: 10.1109/EMBC48229.2022.9871771

Singh, N. K., & Raza, K. (2021). Medical Image Generation Using Generative Adversarial Networks: A Review. *Studies in Computational Intelligence*, *932*, 77–96. DOI: 10.1007/978-981-15-9735-0_5

Singh, P., & Gupta, S. (2022). Unsupervised learning in GANs: The quest for high-quality image synthesis. *IEEE Transactions on Neural Networks and Learning Systems*, *33*(5), 788–802.

Singh, U. (2008). *A History of Ancient and Early Medieval India: From the Stone Age to the 12th Century*. Pearson Education.

Siromoney, G., Chandrasekaran, R., & Chandrasekaran, M. (1983). Machine recognition of Brahmi script. *IEEE Transactions on Systems, Man, and Cybernetics, SMC-13*(4), 648–654. DOI: 10.1109/TSMC.1983.6313155

Skandarani, Y., Jodoin, P. M., & Lalande, A. (2023). GANs for Medical Image Synthesis: An Empirical Study. *Journal of Imaging*, *9*(3), 69. DOI: 10.3390/jimaging9030069 PMID: 36976120

Smith, J., & Lee, R. (2023). Advances in GAN training. *Journal of AI Research*, *58*(3), 145–162.

Subba, J. R. (2008). *Evaluation Of Man And The Modern Society In Sikkim*. Gyan Publishing House. https://books.google.com.my/books?id=q67nHqe4idQC

Suganthi, K. (2021). Review of medical image synthesis using GAN techniques. In *ITM Web of Conferences* (Vol. 37, p. 01005). EDP Sciences.

Tan, H., Liu, X., Li, X., Zhang, Y., & Yin, B. (2019). *Semantics-Enhanced Adversarial Nets for Text-to-Image Synthesis*. https://doi.org/DOI: 10.1109/ICCV.2019.01060

Tan, H., Liu, X., Liu, M., Yin, B., & Li, X. (2021). KT-GAN: Knowledge-Transfer Generative Adversarial Network for Text-to-Image synthesis. *IEEE Transactions on Image Processing*, *30*, 1275–1290. DOI: 10.1109/TIP.2020.3026728 PMID: 33001801

Tan, H., Liu, X., Yin, B., & Li, X. (2023, December). DR-GAN: Distribution Regularization for Text-to-Image Generation. *IEEE Transactions on Neural Networks and Learning Systems*, *34*(12), 10309–10323. DOI: 10.1109/TNNLS.2022.3165573 PMID: 35442894

Ting, D. S. W., Cheung, C. Y.-L., Lim, G., Tan, G. S. W., Quang, N. D., Gan, A., Hamzah, H., Garcia-Franco, R., San Yeo, I. Y., Lee, S. Y., Wong, E. Y. M., Sabanayagam, C., Baskaran, M., Ibrahim, F., Tan, N. C., Finkelstein, E. A., Lamoureux, E. L., Wong, I. Y., Bressler, N. M., & Wong, T. Y. (2017). Development and Validation of a Deep Learning System for Diabetic Retinopathy and Related Eye Diseases Using Retinal Images from Multi-Ethnic Populations with Diabetes. *Journal of the American Medical Association*, *318*(22), 2211–2223. DOI: 10.1001/jama.2017.18152 PMID: 29234807

Tolosana, R., Vera-Rodriguez, R., Fierrez, J., Morales, A., & Ortega-Garcia, J. (2020). Deepfakes and beyond: A survey of face manipulation and fake detection. *Information Fusion*, *64*, 131–148. DOI: 10.1016/j.inffus.2020.06.014

Torfi, A., Fox, E. A., & Reddy, C. K. (2022). Differentially private synthetic medical data generation using convolutional GANs. *Information Sciences*, *586*, 485–500. DOI: 10.1016/j.ins.2021.12.018

Training a neural network on MNIST with Keras | TensorFlow Datasets. (n.d.). Retrieved January 27, 2025, from https://www.tensorflow.org/datasets/keras_example

Tran, N. T., Tran, V. H., Nguyen, N. B., Nguyen, T. K., & Cheung, N. M. (2021). On data augmentation for GAN training. *IEEE Transactions on Image Processing*, *30*, 1882–1897. DOI: 10.1109/TIP.2021.3049346 PMID: 33428571

Tripathi, S., Augustin, A. I., Dunlop, A., Sukumaran, R., Dheer, S., Zavalny, A., Haslam, O., Austin, T., Donchez, J., Tripathi, P. K., & Kim, E. (2022). Recent advances and application of generative adversarial networks in drug discovery, development, and targeting. *Artificial Intelligence in the Life Sciences*, *2*, 100045. DOI: 10.1016/j.ailsci.2022.100045

Vaidya, R., Kumar, P., & Singh, G. (2022). Survey on conditional GANs for image-to-image translation. *Artificial Intelligence Review*, *44*(2), 231–246.

Van Den Oord, A., Kalchbrenner, N., Espeholt, L., Kavukcuoglu, K., Vinyals, O., & Graves, A. (2016). *Conditional Image Generation with PixelCNN Decoders*. https://api.semanticscholar.org/CorpusID:14989939

Vaswani, A., Shazeer, N., Parmar, N., Uszkoreit, J., Jones, L., Gomez, A. N., Kaiser, L., & Polosukhin, I. (2017). Attention is All you Need. *arXiv (Cornell University)*, *30*, 5998–6008. https://arxiv.org/pdf/1706.03762v5

Verdoliva, L. (2020). Media forensics and deepfake detection: Theories, methods, and challenges. *IEEE Signal Processing Magazine*, *37*(3), 42–50. DOI: 10.1109/MSP.2020.2978432

Wang, T. C., Liu, M. Y., Zhu, J. Y., Tao, A., Kautz, J., & Catanzaro, B. (2018). High-resolution image synthesis and semantic manipulation with conditional GANs. *Proceedings of the IEEE Conference on Computer Vision and Pattern Recognition*, 8798-8807. https://doi.org/DOI: 10.1109/CVPR.2018.00917

Wang, X., Sun, L., Chehri, A., & Song, Y. (2023). A review of GAN-based super-resolution reconstruction for optical remote sensing images. *Remote Sensing (Basel)*, *15*(20), 5062. DOI: 10.3390/rs15205062

Wesselkamp, V., Rieck, K., Arp, D., & Quiring, E. (2022, May). Misleading deepfake detection with gan fingerprints. In 2022 IEEE Security and Privacy Workshops (SPW) (pp. 59-65). IEEE.

Wu, Y., Gao, R., Park, J., & Chen, Q. (2020). Future video synthesis with object motion prediction. In *Proceedings of the IEEE/CVF Conference on Computer Vision and Pattern Recognition* (pp. 5539-5548). DOI: 10.1109/CVPR42600.2020.00558

Xu, J., & Liu, Y. (2023). Evolution of GAN architectures for facial recognition and synthesis. *Computer Vision and Image Understanding*, *207*, 102215.

Xu, T., Zhang, P., Huang, Q., Zhang, H., Gan, Z., Huang, X., & He, X. (2018). Attngan: Fine-grained text to image generation with attentional generative adversarial networks. In *Proceedings of the IEEE conference on computer vision and pattern recognition* (pp. 1316-1324).

Yang, H., Ding, W., & Yin, C. (2022). AAE-Dpeak-SC: A novel unsupervised clustering method for space target ISAR images based on adversarial autoencoder and density peak-spectral clustering. *Advances in Space Research*, *70*(5), 1472–1495. DOI: 10.1016/j.asr.2022.05.068

Yang, Q., & Zhang, T. (2022). From GANs to adversarial learning: A new perspective on optimization. *Journal of Optimization Theory and Applications*, *59*(1), 105–120.

Yao, X., Zhang, M., & Wang, Y. (2022). Real-time generation of deepfake content using GANs: Challenges and ethical implications. *Journal of AI Ethics*, *9*(4), 45–60.

You, A., Kim, J. K., Ryu, I. H., & Yoo, T. K. (2022). Application of generative adversarial networks (GAN) for ophthalmology image domains: A survey. *Eye and Vision (London, England)*, *9*(1), 6. DOI: 10.1186/s40662-022-00277-3 PMID: 35109930

Yu, J. W. L., Zhang, W., & Yu, Y. "Seqgan: Sequence generative adversarial nets with policy gradient," p. 2852–2858, 2017.

Yuan, L., Yu, S., Yang, Z., Duan, M., & Li, K. (2023). A data balancing approach based on generative adversarial network. *Future Generation Computer Systems*, *141*, 768–776. DOI: 10.1016/j.future.2022.12.024

Yue, Y., Liu, H., Meng, X., Li, Y., & Du, Y. (2021). Generation of high-precision ground penetrating radar images using improved least square generative adversarial networks. *Remote Sensing (Basel)*, *13*(22), 4590. DOI: 10.3390/rs13224590

Yu, J., Lin, Z., Yang, J., Shen, X., Lu, X., & Huang, T. S. (2018). Generative image inpainting with contextual attention. *Proceedings of the IEEE Conference on Computer Vision and Pattern Recognition (CVPR)*, 5505-5514. https://doi.org/DOI: 10.1109/CVPR.2018.00577

Zhang, H., Xu, T., & Li, H. (2017). *StackGAN: Text to Photo-Realistic Image Synthesis with Stacked Generative Adversarial Networks*. https://doi.org/DOI: 10.1109/ICCV.2017.629

Zhang, W. (2018, June). Generative adversarial nets for information retrieval: Fundamentals and advances. In The 41st International ACM SIGIR Conference on Research & Development in Information Retrieval (pp. 1375-1378).

Zhang, H., Goodfellow, I., Metaxas, D., & Odena, A. (2018). Self-attention generative adversarial networks. *International Conference on Machine Learning*, 7354-7363. https://doi.org/DOI: 10.48550/arXiv.1802.05751

Zhang, H., Goodfellow, I., Metaxas, D., & Odena, A. (2019). Self-attention generative adversarial networks. *Proceedings of the 36th International Conference on Machine Learning*, 7354-7363. https://doi.org/DOI: 10.48550/arXiv.1805.08318

Zhang, H., Xu, T., Li, H., Zhang, S., Wang, X., Huang, X., & Metaxas, D. N. (2019). StackGAN++: Realistic Image Synthesis with Stacked Generative Adversarial Networks. *IEEE Transactions on Pattern Analysis and Machine Intelligence*, *41*(8), 1947–1962. DOI: 10.1109/TPAMI.2018.2856256 PMID: 30010548

Zhang, K., Zhao, X., & Li, H. (2022). Exploring deep learning models for image generation using GANs. *Journal of Artificial Intelligence and Image Processing*, *7*(4), 56–67.

Zhang, L., & Shen, H. (2022). Image-to-image translation with conditional GANs: An investigation. *International Journal of Computer Vision*, *47*(2), 123–139.

Zhang, L., & Zhang, H. (2023). Enhancements in generative adversarial networks for fashion design and clothing generation. *Fashion Technology Journal*, *5*(1), 42–58.

Zhou, T., Li, Q., Lu, H., Cheng, Q., & Zhang, X. (2023). GAN review: Models and medical image fusion applications. *Information Fusion*, *91*, 134–148. DOI: 10.1016/j.inffus.2022.10.017

Zhou, X., & Chen, Y. (2021). Hybrid generative models for image synthesis. *Neural Computation Journal*, *29*(2), 99–121.

Zhu, J. Y., Park, T., Isola, P., & Efros, A. A. (2017). Unpaired image-to-image translation using cycle-consistent adversarial networks. *IEEE International Conference on Computer Vision*, 2223-2232. https://doi.org/DOI: 10.1109/ICCV.2017.244

Zhu, J. Y., Park, T., Isola, P., & Efros, A. A. (2017). Unpaired image-to-image translation using cycle-consistent adversarial networks. In *Proceedings of the IEEE international conference on computer vision* (pp. 2223-2232).

Zhu, M. (2021). A brief analysis of GAN variants on image classification and generation. *Journal of Physics: Conference Series*, *1827*(1), 012165. DOI: 10.1088/1742-6596/1827/1/012165

Zhu, Z., Huang, T., Xu, M., Shi, B., Cheng, W., & Bai, X. (2021). Progressive and aligned pose attention transfer for person image generation. *IEEE Transactions on Pattern Analysis and Machine Intelligence*, *1*, 1. Advance online publication. DOI: 10.1109/TPAMI.2021.3068236 PMID: 33755557

Zühlke, M.-M., & Kudenko, D. (2024). Adversarial Robustness of Neural Networks From the Perspective of Lipschitz Calculus: A Survey. *ACM Computing Surveys*. Advance online publication. DOI: 10.1145/3648351

About the Contributors

Parul Dubey is currently working as an Assistant Professor in Department of Computer Science and Engineering, Symbiosis Institute of Technology, Nagpur Campus, Symbiosis International (Deemed University), Pune, India. Authors highest degree is PhD. in Computer Science and Engineering at C.V. Raman University, Bilaspur, Chhattisgarh, India. She has good academic and research experience in various areas of CSE and IT. She has a teaching experience of four years. She is indulged in research activities which includes book chapters, books, conference papers and journal articles as well. She has 15 Indian published patents and 1 Indian Granted patent. She holds around 45 publications which are part of conferences, scopus and other journals as well. Author is an AWS Certified Cloud Practitioner badge holder, which is a proof for the expertise in AWS. Currently guiding many Engineering students to learn the AWS platform and complete their projects. She has two authored books from international publishers. She has three edited books from international publishers.

Sapana Singh Kshatri, a seasoned scholar with a Ph.D. in Computer Science and Application, specializing in Machine Learning and Data Science from MATS University, Raipur. With an impressive publication record, she has authored papers in SCI and Scopus-indexed journals, UGC-listed papers, and contributed to numerous book chapters. Notably, she has published two books with IGI Global and has two more in the pipeline with Bentham and Wiley. Dr. Kshatri's expertise lies in designing learning frameworks for biomedical and biological data, particularly in disease prediction. With six years of teaching experience in engineering and science colleges, she brings passion and dedication to both education and research. Her impact on the fields of Machine Learning and Data Science is undeniable, marking her as a prominent figure in these evolving domains.

Devanand Bhonsle received his B.E. in Electronics & Telecommunication (ETC) from Pt. Ravishankar Shukla University (PRSU), Raipur in 2004, M. E.

(ETC) from CSVTU, Bhilai in 2008 and Ph.D. (ETC) from CSVTU, Bhilai in 2019. Recently he is working as Sr. Assistant Professor in Shri Shankaracharya Technical Campus (SSTC), Bhilai with experience of 18 years and 7 months. His research area is digital signal and image processing, solar Energy and Internet of Things. He published 50 research papers in various journals and conferences. He authored 8 book chapters and 1 book in reputed publications. He is the reviewer in IEEE access, IET image processing, Electronics letters and IGI global publications. He is the professional member of IFERP.

Bui Thanh Hung, received his M.S. degree and Ph.D. degree from Japan Advanced Institute of Science and Technology (JAIST) in 2010 and in 2013.He was the Director of Data Analysis Artificial Intelligence Laboratory– DAAI Lab, Thu Dau Mot university (2018-2021); Director of Artificial Intelligence Laboratory –AI Lab, Faculty of Information Technology, Ton Duc Thang University (2021-2022).Now He work for Data Science Laboratory, Data Science Department, Faculty of Information Technology, Industrial University of Ho Chi Minh city. He has completed 2 projects, published 2 books, 22 journals, 22 book chapters, 39 International conference papers and 16 domestic conference papers. He had four best paper awards of FICTA 2018, RICE 2020, ICAIAA 2021 and CIEMA 2022. He was selected as Excellent Scientific Researcher of Thu Dau Mot University in 2019, 2020, 2021; and got CIEMA Research Excellence Awards 2022.He is a reviewer of many reputed journals and conferences. He is also a keynote speaker of many domestic and international conferences. His main research interests are Natural Language Processing, Machine Learning, Machine Translation, Text Processing, Data Analytics, Computer Vision and Artificial Intelligence.

***

Minakshi Ashok Ramteke, Assistant Professor, GHRCE, Nagpur, has Completed her Master's Degree from Rashtrasant Tukadoji Maharaj Nagpur University (RTMNU), Nagpur and pursuing PhD in Computer Technology from Rashtrasant Tukadoji Maharaj Nagpur University (RTMNU), Nagpur. She has a total experience of 13 years academics experience. She has published over six research papers in international journals and conferences, all indexed in Scopus. Additionally, she has authored a book and contributed to two book chapters, both of which are also indexed in Scopus.

Pranali Dhawas, an accomplished educator and technologist, is dedicated to teaching and staying updated on emerging technologies. With a wealth of experience, she has made substantial contributions to education, spanning teaching, research, and innovative projects. She has a 6-year professional background, previously affiliated with RTMNU Nagpur University and currently working at G H Raisoni College of Engineering, Nagpur since 2022. Driven by a commitment to knowledge sharing, Prof. Pranali Dhawas has authored and significantly contributed to diverse books and research papers, including topics like Machine Learning, Data Analytics, Big data, Data Pre-processing and Natural Language Processing. Her papers are published in reputable journals like Scopus, reflecting his dedication to expanding knowledge boundaries. Prof. Pranali's skills extend beyond teaching to encompass technical proficiency.

Neha Gautam, a highly qualified expert in machine learning, deep learning, and computer vision, brings over 8 years of experience in the field. She completed her Ph.D. from the prestigious University Malaysia Sarawak, establishing a strong foundation in advanced computing techniques. Dr. Gautam has an impressive academic and professional background, with a record of six SCOPUS-indexed publications and three SCI-indexed publications, showcasing her significant contributions to research. Additionally, she holds a patent, highlighting her commitment to innovation and intellectual property in cutting-edge technology.

Mahendra Kumar Gourisaria is presently working as an Assistant Professor in the School of Computer Engineering at KIIT University, Bhubaneswar, Odisha. He has received his Master degree in Computer Application from Indira Gandhi National Open University, New Delhi and M.Tech in Computer Science and Engineering from Biju Patnaik University of Technology - Rourkela. He is pursuing his Ph.D. from KIIT Deemed to be University. He has an experience of more than 18 years in academia and 7 years in research. He has published more than 20 research papers in different international journals and conferences of repute. He has also served as the organizing committees members of various conferences and workshop. His area of research includes Cloud Computing, Data Mining, Soft Computing and Internet and Web Technology. He is a member of IAENG, UACEE and life member of ISTE, CSI and ISCA.

Keshika Jangde is a PhD pursuing students in Computer Science and Engineering Department at Shri Rawatpura Sarkar University, Raipur (C.G.), working under the supervision of Dr. Ranu Pandey. Keshika completed her undergraduate degree B.E. in Computer Science and Engineering at Institute of Technology Guru Ghasi Das University Bilaspur (C.G.) and her master's degree M.Tech in Computer Technology and Application at the Chhattisgarh Swami Vivekanand Technical University, Bhilai (C.G.) working under the supervision of Dr. Rohit Raza. She is GATE qualified in 2007 and 2008 and also UGC- NET

Qualified in 2018.She is currently working as Assistant professor in Computer Science and Engineering Department at Shri Shakaracharya Institute of Professional Management & Technology, Raipur (C.G.). In her free time, keshika enjoys with her kids and do gardening.

Deepika Roselind Johnson is working as an Assisitant Professor(Senior Grade) in School of Computer Science and Engineering at Vellore Institute of Technology Chennai Campus, Tamilnadu, India. Her area of specialization is Computer Vision, Machine Learning and Internet of Things.

Mangala Madankar is Head of Department of ompurt Science and Engineering. She is having total experience of 20 years, 17 years in the field of academics and 3 years industry experience. She did her PhD in Information Technology from RTMNU, Nagpur. M.E. in Wireless Communication and Computing and B.E. in Computer Engineering, from RTMNU, Nagpur. Her areas of research interests are Natural Language Processing, Data Science, Big Data and Information Retrieval Systems.

Soumya Ranjan Mishra is presently working as an Assistant Professor (II) at the School of Computer Engineering, Kalinga Institute of Industrial Technology (Deemed to be University), Bhubaneswar Odisha, India. He has teaching and research experience of more than 10+ years and Industry experience of 2+ years in IT Companies. He did his doctoral degree Ph.D. in the field of Computer vision and machine learning from the National Institute of Technology, (NIT) Durgapur, WB, India. His research interest includes Image Processing, Intelligent Systems, Human-Computer Interaction, Computer Vision and similar innovative areas. His research contribution includes various proceedings/books which include ASIC Springer series, and publications in reputed conferences, book chapters, and journals indexed in SCI/ESCI.

Hitesh Mohapatra received a B.E. degree in Information Technology from Biju Patnaik University of Technology (BPUT), Odisha in 2006, and the MTech. Degree in CSE from Odisha University of Technology and Research (OUTR) University, Odisha in 2009. He received his PhD in Computer Science & Engineering in 2021 from Veer Surendra Sai University of Technology (VSSUT), Burla, India. He has contributed 35 SCI and Scopus-indexed research papers, 21 international/national conferences and books on Software Engineering Smart City with IoT, and C Programming respectively. He has 15 years of teaching experience both in industry and academia. He has served the research community in various capacities like session chair, technical chair, keynote speaker etc. His research interests include wireless sensor networks, smart city, smart grid and smart water. Currently, he is working as an Associate Professor at the School of Computer Engineering, KIIT (Deemed to be) University, Bhubaneswar, and Odisha.

Manjushree Nayak is currently working as Associate Professor in the Department of Computer Science and Engineering at NIST UNIVERSITY, Berhampur, Odisha. Former working as Assistant Professor in MSIT Department of MATS University Raipur (C.G). She has 18 years' experience of work in academic and Administration. She has completed her Ph.D. in Computer Science in 2020. Her areas of interest are Big data, Machine Learning, Soft computing, Wireless Sensor Network, IOT, Deep learning, Data Analytics. She has published three books. She has published more than 45 papers in National and international journals. Reviewers of many reported journals like Springer, IEEE, IGI global etc. She has delivered more than 5 Keynote/Invited Talks and chaired many Technical Sessions in National and International conferences, Life members in Indian Science congress and International Association of Engineers, Computer Sc. Teacher Association, Science Publishing.

Subhakanta Panda has a strong passion for software engineering and has worked on diverse projects, including Java-based message decryption programs and interactive web applications using HTML, CSS, and JavaScript. He also completed an internship at OctaNet, gaining hands-on experience in web development technologies like HTML, CSS, Java, and SQL. I have also published a chapter on the implementation of an online grocery recommendation system using Python Flask and machine learning. ". He has already worked on the field of IoT and IPV6 and have experience in the field of Machine Learning, Deep Learning using Python. He actively participates in workshops and seminars, including AWS Cloud and AI, to enhance his technical skills.

Siji Thomas, is an Engineer by profession, currently accomplishing her career with practical understanding of Industrial Management System. Her great vocation in technical publications led to develop herself as a professional Independent Researcher and has now specialized in Robotics, ML, DIP and IoT, where she has made her contributions. Her milestones in research career proven her expertise in reviewing technical chapters. She is the recipient of the National Special Jury Award for developing a Smart Biomedical Device. She has proven her professional dedication in various roles, including Assistant Professor, Calibration Engineer, L&D, US Online Instructor etc. She is proficient in Circuit Design and her functional competencies include MATLAB & LabVIEW, thereby leveraging her focus in driving technological innovations. Ms. Thomas earned her Master's degree in Communication Engg. from KTU University, India, and Bachelor's degree in Instrumentation Engg. from MG University, India.

Index

H

I

M

N

O

www.ingramcontent.com/pod-product-compliance
Lightning Source LLC
LaVergne TN
LVHW081330021026
849161LV00008BA/1927
9798369371527